ANTIMICROBIAL SUSCEPTIBILITY TESTING

Critical Issues for the 90s

ADVANCES IN EXPERIMENTAL MEDICINE AND BIOLOGY

Recent Volumes in this Series

Volume 342
CORONAVIRUSES: Molecular Biology and Virus–Host Interactions
Edited by Hubert Laude and Jean-François Vautherot

Volume 343
CURRENT DIRECTIONS IN INSULIN-LIKE GROWTH FACTOR RESEARCH
Edited by Derek LeRoith and Mohan K. Raizada

Volume 344
MECHANISMS OF PLATELET ACTIVATION AND CONTROL
Edited by Kalwant S. Authi, Steve P. Watson, and Vijay V. Kakkar

Volume 345
OXYGEN TRANSPORT TO TISSUE XV
Edited by Peter Vaupel, Rolf Zander, and Duane F. Bruley

Volume 346
INTERACTIVE PHENOMENA IN THE CARDIAC SYSTEM
Edited by Samuel Sideman and Rafael Beyar

Volume 347
IMMUNOBIOLOGY OF PROTEINS AND PEPTIDES VII: Unwanted Immune Responses
Edited by M. Zouhair Atassi

Volume 348
ADVANCES IN NUTRITION AND CANCER
Edited by Vincenzo Zappia, Marco Salvatore, and Fulvio Della Ragione

Volume 349
ANTIMICROBIAL SUSCEPTIBILITY TESTING: Critical Issues for the 90s
Edited by James A. Poupard, Lori R. Walsh, and Bruce Kleger

Volume 350
LACRIMAL GLAND, TEAR FILM, AND DRY EYE SYNDROMES: Basic Science and Clinical Relevance
Edited by David A. Sullivan

ANTIMICROBIAL SUSCEPTIBILITY TESTING

Critical Issues for the 90s

Edited by

James A. Poupard
SmithKline Beecham Pharmaceuticals
King of Prussia, Pennsylvania

Lori R. Walsh
Abington Memorial Hospital
Abington, Pennsylvania

and

Bruce Kleger
Pennsylvania Department of Health
Lionville, Pennsylvania

PLENUM PRESS • NEW YORK AND LONDON

Library of Congress Cataloging-in-Publication Data

Antimicrobial susceptibility testing : critical issues for the 90s / edited by James A. Poupard, Lori R. Walsh, and Bruce Kelger.
p. cm. -- (Advances in experimental medicine and biology ; v. 349)
"Proceedings of an Eastern Pennsylvania Branch of the American Society for Microbiology Symposium on Antimicrobial Susceptibility Testing : Critical Issues for the 90s, held November 14-15, 1991, in Philadelphia, Pennsylvania"--T.p. verso.
Includes bibliographical references and index.
ISBN 0-306-44673-1
1. Microbial sensitivity tests--Congresses. I. Poupard, James A. II. Walsh, Lori R. III. Kleger, Bruce. IV. Eastern Pennsylvania Branch of the American Society for Microbiology Symposium on Antimicrobial Susceptibility Testing: Critical Issues for the 90s (1991 : Philadelphia, Pa.) V. Series.
[DNLM: 1. Microbial Sensitivity Tests--congresses. W1 AD559 v.349 1993 / QW 25 A631 1993]
QR69.A57A58 1993
616'.01--dc20
DNLM/DLC
for Library of Congress 94-687
CIP

Proceedings of an Eastern Pennsylvania Branch of the American Society for Microbiology symposium on Antimicrobial Susceptibility Testing: Critical Issues for the 90s, held November 14–15, 1991, in Philadelphia, Pennsylvania

ISBN 0-306-44673-1

A Division of Plenum Publishing Corporation
233 Spring Street, New York, N.Y. 10013

Printed in the United States of America

SYMPOSIUM COMMITTEE

Program Chairman
James A. Poupard, Ph.D.
SmithKline Beecham Pharmaceuticals

Program Co-Chairpersons
Bruce Kleger, Dr. P.H.
PA Department of Health

Lori R. Walsh, M.S.
Abington Memorial Hospital

Committee Members

Carl Abramson, Ph.D.
PA College of Podiatric Medicine

Josephine Bartola, J.D.
PA Department of Health

Kenneth R. Cundy, Ph.D.
Temple University School of Medicine

Anna Feldman-Rosen, M.S.
Thomas Jefferson University

Olarae Giger, Ph.D.
Episcopal Hospital

Diane Halstead, Ph.D.
HealthEast Laboratories

Linda A. Miller, Ph.D.
Holy Redeemer Hospital & Medical Center

Joel Mortensen, Ph.D.
St. Christopher's Hospital for Children

James E. Prier, Ph.D.
Philadelphia College of Osteopathic Medicine

Stephen F. Rittenhouse, M.A.
SmithKline Beecham Pharmaceuticals

Donald D. Stieritz, Ph.D.
Hahnemann University

GENEROUS FINANCIAL SUPPORT FOR THIS SYMPOSIUM HAS BEEN PROVIDED BY:

AB BIODISK NORTH AMERICA, INC.

ALAMAR

API

HOECHST-ROUSSEL PHARMACEUTICALS, INC.

MERCK SHARP & DOHME

MICROSCAN DIVISION-BAXTER HEALTHCARE SYSTEM

ORTHO PHARMACEUTICAL CORP.

ROERIG DIVISION - PFIZER INC.

SMITHKLINE BEECHAM PHARMACEUTICALS

VITEK SYSTEMS

PREFACE

The papers assembled in this collection comprise a majority of the oral presentations as well as several poster presentations given at the 22nd Annual Symposium arranged by the Eastern Pennsylvania Branch of the American Society for Microbiology.

The symposium would not be possible without the generous support of the many sponsors (see sponsor list) or without the concerted effort of all the Committee members. This Symposium series has evolved into an annual Eastern Pennsylvania Branch ASM event that attracts participants from a wide geographic area. It should be noted that one of the hallmarks of these symposia involves interaction between the presenters and those in attendance. Several authors have altered their manuscripts based on comments by the participants. Therefore, the manuscript that follows should be viewed as a group effort of both the participants and presenters.

James Poupard

Lori Walsh

Bruce Kleger

CONTENTS

INTRODUCTION

The purpose of the symposium and the works collected in this book is to focus attention on the issues relating to antimicrobial susceptibility testing that face the clinical microbiologist in the last decade of the current century. The accent is placed on issues rather than actual methodology. The organizers felt that there was a wealth of information on the test systems available in the contemporary literature or as information that can be obtained from the instrument manufacturers. The only methods that have received specific attention are those that have been recently introduced or significantly modified in recent years. Of course the issues of the '90's are based on earlier developments, but because of many factors, the whole field appears to be more sharply defined in the '90s than in the previous decade. Many of these factors are related to increased emphasis on cost containment or are related to valuable lessons learned from the proliferation of automated systems that occurred in the 1980s.

Because of this changing perspective, the first section of this work is dedicated to a brief chronological review of the history and evolution of contemporary methods of antimicrobial susceptibility testing. This was felt to be important because, although many of the contemporary issues are unique to the '90s, they have definite roots in the past. Although a sense of the past is not necessary to comprehend the present methodologies, to ignore the past is foolish. One cannot get a full perspective of this contemporary field without understanding the problems and issues of the past. This is particularly true of antimicrobial susceptibility testing for many reasons. It is difficult to comprehend many of the unique aspects of this field without a brief survey of how the field became so complex in a relatively short span of time. Especially as it relates to the automated test systems. Automation is an area that is somewhat peculiar to susceptibility testing as it is practiced in the United States. Many of the issues that were faced in the United States in the '70s and '80s are now contemporary issues in Europe and other parts of the industrialized world. It is hoped that the collection of works assembled by the top practitioners in the field will be a valuable source of information to microbiologists in other countries, and that this "snapshot" of some of the contemporary issues will assist those microbiologists who are newly exposed to some of the automated test systems. This may enable them to bypass some of the "errors" before plunging into their assessment of the need and relevance of the various susceptibility test systems now available in North America. As the entire field of production and dissemination of antimicrobial agents becomes increasingly transnational, the peculiar nature of the U.S. accent on breakpoints, automated methodologies and external agency "mandatory" standardization may become a more international issue then it is today. One of the goals of the current work is to help define rather than resolve many of these issues. Each invited contributor to this work was chosen with this goal in mind.

Regardless of how thorough the organizing committee was in selecting experts, the real experts on the issues are the practicing microbiologists who face many of the consequences of susceptibility testing policies in their everyday work. The organizers devised a way to permit each of the approximately 250 attendees of the symposium to participate in its content. A portion of the symposium was dedicated to a session called, "Contemporary Issues: Point and Counterpoint." This was a unique session in which the actual issues were presented to the attendees for their comments. Questionnaires were given to each attendee asking for information and opinions on many issues as well as providing an area for free text to permit attendees to express what they felt were significant issues that may not have been anticipated by the experts. These comments were collated and are presented in this work as a section dedicated to "Key Contemporary Issues." Some of these are rather complex, and it has taken some time to better define these points and to get consensus on the issues. The editors apologize to the readers and the authors for the length of time from the actual symposium to the publication of this work. It is hoped that the end product justified the delay and we would like to thank the contributors for their patience.

THE EVOLUTION OF ANTIMICROBIAL SUSCEPTIBILITY TESTING METHODS

James A. Poupard[1], Stephen F. Rittenhouse[1]
and Lori R. Walsh[2]

SmithKline Beecham[1]
King of Prussia, PA

Abington Memorial Hospital[2]
Abington, PA

INTRODUCTION

It is reasonable to conclude that antibiotic testing did not precede the concept of antibiotics, therefore, it is logical to start the review with penicillin and the work of Alexander Fleming. It should be noted that the history of antibiotic testing is no different than other topics in the history of science, there are always relevant predecessors to whatever event is chosen as the starting point for the subject. It would be negligent to fail to mention the work of Paul Ehrlich as well as several early observations that seem worthy of note before focusing on the evolution of the events leading to the contemporary field of antimicrobial susceptibility testing.

There is a brief review of the subject by Lechevalier and Solotorobsky.[1] Anyone interested in the early observations that preceded the work of Fleming should consult this 1965 publication. The published works of Pasteur, Koch and especially Paul Ehrlich contain many references to antibiosis. Along with these observations are descriptions of laboratory methodologies that support the concept of antibiosis. This methodology essentially consisted of measuring activity by employing: (i) test tubes containing broth (but technically not a MIC determination), (ii) a loss of motility, or (iii) animal protection studies. Although these observations were significant, they did not establish practical methodologies that were predictive of clinical outcome with the possible exception of animal studies conducted by Ehrlich. Since this review focuses on *in-vitro* testing, the importance of the work mentioned above is dutifully noted. However, before leaving this subject, recognition should be given to two early events that illustrate the types of laboratory observations that were made in the early years following the establishment of bacteriology as a separate branch of science. In 1874, William Roberts observed that liquid medium in which the mold Penicillium glaucum was growing, could not be easily contaminated with bacteria.[2] In 1876, John Tyndall

Antimicrobial Susceptibility Testing, Edited by
J.A. Poupard *et al.*, Plenum Press, New York, 1994

as noted by Lechevalier and Solotorobsky[1] made a similar observation that broth supported the growth of either bacteria or mold, but rarely both. These examples illustrate that in the early years of bacteriology, laboratory workers were observing the effects of inhibiting substances as well as inhibition between organisms. It also illustrates that broth was employed to establish the inhibiting effect of these metabolic products on the growth of bacteria in the laboratory. These and other observations were essentially isolated events that did not lead to the establishment of an organized series of continuing studies.

It is almost an understatement to note that Paul Ehrlich was the first bacteriologist to devote considerable effort focused on discovering therapeutic cures for infectious diseases employing chemical agents. As director of the Konigliches Institute, he stated that the task of this new institute was to find "specific chemotherapy of infectious diseases" by employing products generated in the laboratory rather than by use of substances found in the serum of humans or animals.[1] His pioneering work in chemotherapy established this subject as a separate field of scientific inquiry, however, his techniques tended to be quite organism-specific, and relied heavily on *in-vivo* animal studies. It was 20 years later that work on penicillin began the series of events that the modern day concept of *in-vitro* testing became established. This chain of events formed the basis of modern antimicrobial susceptibility testing methodology. This work was initiated by Alexander Fleming and his contemporaries in the 1920's.

Alexander Fleming made his initial observation on the inhibitory effect of what eventually became known as penicillin on solid media by observing an area of growth inhibition of staphylococcal colonies adjacent to a Penicillium contaminated agar plate. For purposes of the current review, this event will be taken as the starting point for the evolution of contemporary susceptibility testing methodologies.

It is important to state the sources of information for the next three sections of this review. A search of the literature for authoritative surveys on the evolution of antimicrobial susceptibility testing revealed a great variety of works by many authors. Most of these reviews are organized by methodology rather than in a confluent chronological order. The present authors felt that for analysis purposes, a better sense of how the contemporary field of susceptibility testing developed could be obtained by creation of a master chronology that could then be divided into time periods with certain landmark events chosen to represent particular periods of time. By employing this method, a series of short analyses could be made for each of the time blocks.

To accomplish this goal, five major works were selected that, in the judgment of the authors, demonstrated particularly sound and well-researched material. These five works were used to create a chronological table of significant events in the history of the subject. The chronology is presented at the end of this review. The following are listed in the chronology: The date the method was published, the reference source for each method, the primary investigator or author(s) of the paper describing the method, and a brief description of the methodology. This chronology begins in 1924 and continues through 1988. In the text that follows, reference is only made to the investigator and date. If one is interested in more information concerning any of the methods described in the text, the chronology can be consulted with the corresponding reference source. It should be noted that only certain events were selected from the chronology for inclusion in the text. Although this may not be a conventional reference method, the authors believe that this system contributes to the flow of information presented in the text while permitting anyone who wants more information to consult this specific source and reference. The five sources used in this master chronology are the following: P.F. Wheat and R.C. Spencer,[3] H.D. Isenberg,[4] E. Wilson, D.A. Henry and J.A. Smith,[5] K.G. Stubs and K. Wicher,[6] and L. Walsh.[7]

From Fleming to WWII

In addition to the discovery of penicillin, Alexander Fleming made two significant contributions to the field of antibiotic testing in the 1920's. In 1924, he introduced the use of the ditch plate technique for evaluating the antimicrobic qualities of a test solution. This involved seeding the agar with a test organism, cutting a ditch in the agar, placing the test solution in the ditch and observing for areas of inhibition around the ditch after appropriate incubation. It should be noted that in 1929, this technique was modified by Reddish, who cut wells into the agar and filled the wells with the test solution. In 1929, Fleming introduced a broth dilution method that could be considered a definite forerunner to contemporary MIC methodology.

Four investigators made significant contributions in 1940. Pope introduced the use of absorbent paper that had been impregnated with the test antibacterial substance for evaluating susceptibility. Schmith and Reymann introduced the first agar dilution susceptibility test method that year, and Gardner introduced a broth dilution technique in which he used morphological changes in the organism as an endpoint, instead of relying strictly on turbidity. This later method could be considered a forerunner to the current day methods used to evaluate post antibiotic effect. In 1941, Abraham et. al. introduced the use of a cylinder placed into the agar followed by placing the antimicrobial agents in the cylinder. This permitted the agent to diffuse into the agar. These cylinders were known by several names including the Oxford Cup, the Fish Spine, and the Heatly Cup. In the U.S., these cylinders often were referred to as penicillin cylinders, and the penicillin cylinder method for evaluating various antimicrobial substances became a popular method of testing. The next year, Alexander Fleming introduced another modification of the broth dilution method by using pH as an indicator instead of turbidity. It is interesting to note that other investigators modified this technique using variations of the pH method, but in general, this met with little success. The first truly practical application of using pH as an indicator did not come about until quite recently with the introduction of the Alamar System. This system is raising considerable interest in this country as a potentially new and practical method of performing susceptibility testing employing pH as a means for reading endpoint values.

In 1945 much progress was made in the use of antibiotic impregnated disks for evaluating susceptibility of an organism to a drug. Mohs introduced a "radial streak disk method" involving the comparison of a test organism with a sensitive control on the same plate. This method was a forerunner to the Stokes method used in many parts of Europe to the present time. Other investigators such as Copeland, and Morley also introduced technical modifications of the disk method this year. Frisk published his work on the incorporation of penicillin into the agar for evaluating susceptibility of S. pneumoniae in 1945.

Therefore, by the end of WWII, the concept that penicillin was a drug for treating many infectious diseases became well established. The tube dilution methodology, agar diffusion methods employing disks, as well as several agar dilution techniques all became accepted as methods for accurately predicting penicillin susceptibility as this period ended.

WWII to 1975

This period began with the establishment of penicillin as a definite therapeutic agent and witnessed the introduction of new antimicrobial agents designed to treat both gram positive and gram negative infections. As this period progressed, and the number of drugs increased, the need to predict susceptibility and select those agents best suited to treat a particular infection became greater. The need to determine susceptibility or resistance to multiple drugs using practical methodology stimulated innovative approaches to this problem. Garrett in 1946 introduced a multiple replication device to perform large numbers of agar dilution tests, and introduced the concept of critical dilutions as opposed to large numbers of serial dilutions for evaluating susceptibility.

In 1947 several workers introduced modifications of the disk technique. Hoyt and Levine introduced penicillin tablets as an alternate method to the use of penicillin disks, and in that year Bondi and workers introduced the now standard 6.5 mm disk. In 1952, Gould and Bowie did extensive work on a technique using disk diffusion of varying concentration disks on the same plate containing a control organism to judge susceptibility or resistance. Three years later in 1955, Stokes introduced the method that bears her name. This method became well known and was widely used throughout Europe. In 1956, Kirby et. al. introduced the single disk method of performing susceptibility testing. Steers, Foltz, and Graves introduced an agar dilution method using a replicating device that is still used as a means of susceptibility testing with several other applications. In 1959, Bauer and coworkers introduced the high potency single disk method that, with additional work, eventually became known as the Bauer Kirby Method of susceptibility testing. Almost simultaneously, Anderson and Troyanosky introduced another standardized disk technique employing low-potency disks. In 1963 Tolhurst, Buckle and Williams made significant progress in modifying the full tube dilution technique by focusing on dilutions surrounding critical breakpoints rather than using the cumbersome full range of dilutions to accommodate the testing of many drugs against a particular clinical isolate.

Starting in 1959, the need to standardize susceptibility testing became apparent, and several organizations and investigators began addressing this critical issue. In 1959 and 1960 Ericsson and Steers were performing studies to evaluate the various methods that were in use at that time. In 1961, the World Health Organization addressed the issue of standardization of antimicrobial susceptibility test methodology. This was followed in 1964 with Isenberg performing a comparison of susceptibility testing methodologies used in the United States. Also in that year, Truant and coworkers made significant attempts to standardize tube dilution MIC methodology. In 1966 significant progress in standardization of the disk method occurred when Bauer, Kirby, and coworkers published their attempt to firmly establish the disk diffusion technique as a practical method of testing with broad application for the routine hospital clinical laboratory. In 1971, Ericsson and Sherris carried out an international collaborative study to evaluate susceptibility testing methods. In 1972, Stokes and Waterworth made a major contribution in their attempt to standardize susceptibility testing in the U.K. In 1975, the National Committee for Clinical Laboratory Standards (NCCLS) published Antimicrobial Susceptibility Testing Standards. This initiated a series of events which brought many of the earlier attempts to standardize susceptibility testing methodology on a national basis into focus and for the first time provided a practical and accurate method that could be adopted by all clinical laboratories in the U.S.

1975 to Present - Automated Systems

By 1975, the concept of antimicrobial susceptibility testing had become an intricate part of contemporary clinical microbiology. Starting in 1970, clinical microbiology became established as a separate and identifiable discipline within the broader context of medical microbiology.[8] Antimicrobial susceptibility testing played a dominant role in establishing the need for microbiologists who specialized in specimen selection and processing, organism identification, and methods of performing susceptibility testing. The number of antimicrobial agents were proliferating and there was a perceived need for a more sophisticated methodology then that provided by the standard disk diffusion technique. These factors were combined with such perceived notions that reporting results as susceptible, intermediate or resistant, were inferior to methodologies that would result in a MIC endpoint. One additional factor should be noted. Other areas of clinical laboratory science were becoming increasingly automated during this time, and this process was producing significant reductions in cost per test along with significant reduction in actual technologist time per test. The perception that microbiology departments were less progressive because they were so work-intensive became a factor that should not be ignored. Automation appeared to be a dominate factor necessary to modernize the microbiology department.

The success of the BacTec system in automating blood culturing appeared to add validity to the concept that automation could do for clinical microbiology what other instrumentation had done for the clinical chemistry laboratory. An additional element that played a significant role during this time was the proliferation of electronic data processing systems which were permeating all major clinical laboratory departments. The success of computers in resolving many complexities of laboratory data processing was a significant factor in creating the perception that clinical microbiology had to adapt to this necessary modernization. Unfortunately the disk diffusion method did not lend itself to computerization. Automated antimicrobial test systems appeared to be the solution to many of these perceived problems.

This first automated test system to fulfill these perceived needs to bring the microbiology laboratory in line with departments in other areas of laboratory medicine was initiated in 1974.[9] By 1975, interest in this new rapid system became focused on what role this new instrument would play in clinical microbiology laboratories. This system was the Autobac disk elution system introduced by Pfizer Diagnostics. It was a system that afforded machine read results and same day turn around time from the time of inoculation. It had the potential of revolutionizing the whole field of antimicrobial susceptibility testing, depending on how well it was accepted by many reluctant microbiologists. Although the system had several problems, it must be credited as establishing the first automated rapid susceptibility test system to get the attention of the clinical microbiology community in the United States. One of the main drawbacks of the initial system is that it did not have a built-in organism identification element. This caused considerable problems in issuing final susceptibility results without an accompanying organism identification for the report. There was also significant problems in fitting the whole process into the average work day, and in many laboratories, the final result was held until the next day when the identification portion was available, and technologists could process the results that became available the previous evening. This tended to undermine the rapid aspect of the instrument.

The second automated system to gain acceptance was the Abbott MS-2 System, introduced in 1977.[10] This system was a four hour system and included an organism identification portion. This was also a disk elution system and generated calculated MICs. This was shortly followed by the AMS System introduced by McDonnell Douglas Corporation, also in 1977.[11] This was the forerunner of what is known today as the Vitek System. This system utilized dehydrated reagents in sealed plastic cards and contained separate cards for susceptibility testing and organism identification.

Although by no means were these three systems common by 1977, all three had been introduced, and the basic elements of the automated technology that would dominate the next two decades were beginning to have an impact. This impact had its effect on actual usage and the resulting debates on a whole new set of issues that were novel to clinical microbiologists. It is noteworthy that, 1977 was also a significant year for the field of susceptibility testing because of the introduction of another technological innovation that would play a prominent role in future automated susceptibility testing methodology. The introduction of standardized microtiter trays to perform susceptibility testing by Thornsberry and co-workers was a major factor in stimulating the technological developments that offered a significant challenge to the disk diffusion method. It was a method that was commercialized first as frozen panels and later as lyophilized panels, based on a modification introduced in 1978 by Phillips. These panels could be used to generate MICs as well as organism identification from the same inoculum. Automating the reading process for these commercially produced microtiter panels was the next logical step in the development of automated test systems. Like all other technological advances that serve a practical need, this new technology rapidly moved in many different directions.

The application of this technology lead to the development of such test systems as Micro-Media Systems, Sensititre, BBL Sceptor, and a whole line of products by MicroScan which started as a basic microtitre system, and developed into more complex modifications such as the TouchScan, AutoScan, and the current MicroScan WalkAway System.

Space does not permit elaboration on the many systems that were initiated in the late 1970's and throughout the 1980's. For a more detailed account, see the excellent review by Piddock.[12] Suffice it to say that although the field had a relatively slow start by the 1990's the whole field of antimicrobial susceptibility testing appears to be equally divided between those laboratories that employ a fully automated system, those laboratories using some form of microtitre technology and an equal portion using the disk diffusion method as their primary test system.

The presence of these systems for susceptibility testing, combined with rapid organism identification and increased integration with the electronic data processing technology that now permeates clinical laboratory departments has had a significant impact on the practice of clinical microbiology in the United States. It is understandable that these changes have created significant issues that are currently being debated. Many of these key issues will be described in other sections of this book.

As we enter the last decade of this century, the field of susceptibility testing in the U.S. appears to be going through some significant changes. The more complex and expensive automated systems appear to be dominated by two manufacturers, Vitek and MicroScan, with a few other systems that appear to serve specific needs. The disk diffusion methodology has certainly maintained its utility in spite of much criticism over the last 20 years. Disk diffusion techniques have continued to demonstrate certain characteristics such as being an economical system, and one that offers maximum flexibility when compared to all of the other test systems today. Between these two models, the fully automated and the disk diffusion systems, there exists a wide range of semi-automated systems that produce accurate results and give the user the option of generating an MIC or a breakpoint based susceptible, intermediate, or resistant result.

A trend that seems to be gaining popularity in the early '90's is the development of new and innovative test systems such as the E-test that fill a definite need for generating accurate MIC results for fastidious organisms to be used for off-line testing in areas where the fully automated test systems are deficient. Some of these newer systems will be dealt with separately in this book and will not be discussed at this time.

In summary, as the '90's unfold, the one thing that is certain is that it will be an interesting decade for developments in antimicrobial susceptibility testing. Although, the U.S. seems to have adopted automated test systems more rapidly than the rest of the world, the U.S. should serve as a test model for other countries interested in what direction susceptibility testing should follow. This is particularly true because of the existence of large numbers of laboratories using one of the three levels of automated/non-automated systems in the U.S. With the spirit of reduced health care spending, it will certainly be interesting to speculate whether the remainder of the '90's will continue the balance between the three levels of automation in the U.S., or if one of the three general levels will become dominant. Since there does not appear to be an innovative alternative to the current methods, it is probably safe to assume that for the immediate future the current methodologies will be constantly adapted to resolve the current problems and better serve the needs of their users. The introduction of the newer methodologies to serve very specific functions and to supplement the current systems will certainly be an area that will receive considerable attention in the coming years. The current need for producing rapid results for such classically difficult to test organisms as Mycobacterium tuberculosis will probably lead to significant innovations before the end of this century. If history repeats itself, now that the need has been defined, the technology is bound to follow, and the solution will probably "spill-over" to resolve many of the drawbacks that all systems seem to have, thus creating an entirely new batch of issues for future generations of clinical microbiologists to debate.

A CHRONOLOGY OF THE EVOLUTION OF ANTIMICROBIAL SUSCEPTIBILITY TESTING METHODOLOGY 1924 to 1988

DATE	SOURCE*	INVESTIGATOR	METHOD
1924	1	Fleming	Ditch Plate
1929	1	Reddish	Cutting Wells in Agar
	1	Fleming	Broth Dilution Method
1940	1	Pope	Absorbent paper to carry antibacterial
	1	Schmith & Reymann	Agar Dilution AST (first) Method
	5	Gardner	Broth Dilution Tech. (morph changes)
1941	1	Abraham et al.	Cylinder plate ("Oxford Cup", "Fish spine", "Heatly Cup")
1942	1	Fleming	Broth dilution by PH change
	1	Rammelkamp & Maxon	Broth Dilution Tech.
1943	1	Schmidt & Sesler	Broth Dilution Tech.
1944	1	Vincent & Vincent	Filter paper discs with penicillin for assaying.
	5	Garrod & Heatly	Agar Diffusion Method for estimating penicillin in plasma
1945	1	Mohs	"Radial streak" disk compare with a sensitive control
	1	Copeland	Disk Diffusion
	1	Morley	Disk Diffusion
	1	Spink & Ferris	Broth Dilution Tech.
	1	Frisk	Penicillin into agar - Susc. of S. pneu.
1946	1	Buggs et al.	Broth Dilution Tech.

	1	Garret	Multiple replica devices for agar dilution - changed from serial dilution to critical dilution
	5	Duguid	Dilution (liquid) - morph change
1947	1	Kokko	Disk Diffusion
	1	Kolmer	Disk Diffusion
	1	Hoyt & Levine	Penicillin tablets
	1	Bondi et al	Describe standard 6.5 mm disk used today.
1950	1	Frank, Wilcox & Finland	Used agar dilution for many antibiotic/organism combinations
1951	1	Jackson & Finland	Studies comparing agar diffusion and dilution tech.
	5	Waisbren et al.	Described tube dilution MIC
1952	1	Gould & Bowie	U.K. tech using disk diffusion of varying conc. disks with control organisms for comparison.
	5	Szybalski	Described agar dilution methods
1954	1	Ericsson et al	Testing Method
	3	Schneirson	Rapid disk tube method
1955	1	Stokes	Concent. disk compared on same plate to clinical isolate and known organism (U.K.)
1956	5	Kirby et al	Single disk method
1959	1	Steers, Foltz & Graves	Simplified agar dilution replica device and breakpoint concept.
	5	Bauer et al.	High potency single disk method
	1	Ericsson et al	Testing Method

(continued)

A CHRONOLOGY OF THE EVOLUTION OF ANTIMICROBIAL SUSCEPTIBILITY TESTING METHODOLOGY 1924 to 1988 (Continued)

1960	5	Ericsson	Modif. of agar diffusion using 30 min. pre diffusion.
	1	Steers et al.	Comparison studies of agar diffusion & dilution tech.
	5	Anderson & Troyanosky	Single low potency disk
1961	1	World Health Organization	Standardized AST Method
1963	1	Tolhust, Buckle & Williams	Estab. critical breakpoints for dilution
1964	2	Isenberg	Comparison nationwide AST using disks
	5	Truant et al	Standard tube dilution MIC
1966	1	Bauer et al.	Standardized disk diffusion technique
1967	2	Isenberg	Exp. with altering nutrient supplements in diluents.
1971	5	Ericsson & Sherris	Authored an International Collaborative Study
1972	1	Stokes & Waterworth	Standardized AST methods in the U.K.
	1	Rolinson & Russell	Filter paper strips with known drug conc.
1973	1	Haltalin, Markley & Woodman	Critical breakpoints for agar dilution
	3	Wilkins & Theil	Modified Broth Disk Method
1975	1	Shafi	Filter paper pads with spec. drug conc.
	1	NCCLS	AST standards
1976	4	Lorian et al.	5 hour disk susc. test
1977	1	Thornsberry et al.	Standardized microdilution trays

1978	1	Philips et al.	Dehydrated microtiter MIC panels
1980	2	Krogstad & Moellering	Checkerboard studies time kill curves
1981	1	Swedish Reference Group	Standardized AST method in Sweden
1984	1	Deutsches Institut fur Normung.	Standardized ST method in Germany
	1	Snell, Danvers & Gardner	Tablets with spec. drug conc.
1985	1	NCCLS	Established breakpoints
	1	ECCLS	Established breakpoints
1987	1	Doern	Flurogenic microdilution panels
1988	1	WPOBSFAC**	Established breakpoints

*Source: 1. P.F. Wheat & R.C. Spencer 1988 (3), 2. H.D. Isenberg 1988 (4), 3. E. Wilson, D.A. Henry and J.A. Smith 1990 (5), 4. K.G. Stubs and K. Wicher 1977 (6), and 5. L. Walsh 1990 (7).

**WPOBSFAC: Working Party of the British Society for Antimicrobial Chemotherapy.

REFERENCES

1. Lechevalier, H.A. and Solotorvsky M. 1965 Three Centuries of Microbiology. McGraw-Hill Book Co., NY.

2. Roberts, W. Studies on biogenesis. Phil. Trans. Royal Soc. London, 164:466, 1874.

3. Wheat, P.F. and Spencer, R.C. 1988. The evolution of *in-vitro* antimicrobial susceptibility techniques J. of Antimicrob. Chemother. 22:579-582.

4. Isenberg, H.D. 1988. Antimicrobial susceptibility testing: A critical evaluation. J. of Antimicrob. Chemother. 22 suppl. A. 73-86.

5. Wilson, E., Henry, D.A. and Smith, J.A. 1990. Disk elution method for MICs and MBCs. Antimicrobial Agents & Chemo. 34:2128-2132.

6. Stubs, K.G. and Wicher, K. 1977. Laboratory evaluation of an automated antimicrobial susceptibility system. Am. J. Clin. Path. 68:769-777.

7. Walsh, L. 1991. An International Survey of Antimicrobial Susceptibility Testing Methods Employed in the Clinical Microbiology Laboratory, Medical Col. of PA., Univ. Microfilm International, Ann Arbor, Mich.

8. Pouaprd, J.A. 1982. A Comparison and Evaluation of Relevance of Clinical Microbiology Master's Degree Programs in the U.S. Univ. of PA., Univ. Microfilms International, Ann Arbor, Mich.

9. McKie, J., Borovoy, R., Dooley, J., Evanega, G. Mendoza, G., Meyer, F., Moody, M., Packer, D., Praglin, J., and Smith, H. 1974. Autobac I - A three hour automated antimicrobial susceptibility system II Microbiological Studies. In Automation in Microbiology & Immunology. ed., Heden C. and Illeni, T. N.Y., John Wiley.

10 Spencer, H.J., Stockert, J., Welaj, P., Wilborn, R., and Price, B. 1977. Automated antimicrobial susceptibility testing with the MS-2 system. In Rapid Methods and Automation in Microbiology ed. Johnston, H.H. and Newsom, SWB. pp 262-263.

11. Aldridge, C., Jones, P.W., Gibbson, S. Lanham, J. Myer, M., Vannest, R., and Charles, R. 1977. Automated microbiological detection/identification system. J. of Clin. Micro. 6:406-413.

12. Piddock, L.J.V. 1990 Techniques used for the determination of antimicrobial resistance and sensitivity in bacteria. J. of Applied Bact. 68:307-318.

ANTIMICROBIAL SUSCEPTIBILITY TESTS:

TESTING METHODS AND INTERPRETIVE PROBLEMS

Patrick R. Murray

Washington University School of Medicine
Saint Louis, MO

INTRODUCTION

Infectious diseases are a leading cause of morbidity and mortality in hospitalized patients. This fact places a tremendous burden on the clinical microbiology laboratory to diagnose rapidly the etiologic agent(s) responsible for a patient's infection and provide therapeutic guidance for eradication of the organism(s). Laboratories are asked not only to perform these tasks with precision and rapidity, but also in a cost-effective manner in an era of increasing emphasis on reduction of laboratory expenses. Physicians will pose the question - How accurately can antimicrobial susceptibility tests predict clinical outcome? At the same time, the microbiologists ask - What method should we use for performing susceptibility tests? What is the most accurate method? What is the most cost - effective method? What are the limitations of the *in vitro* tests? I will try to address these questions in the following discussion.

CLINICAL VALUE OF IN VITRO SUSCEPTIBILITY TESTS

The limitations of antimicrobial susceptibility tests can be appreciated by remembering that these tests are well - standardized assessments of only one aspect of a very complicated clinical scenario. The standards published by the National Committee on Clinical Laboratory Standards (NCCLS)[8,9] provide specific guidelines for the performance of these tests and, if the guidelines are followed precisely, the results are remarkably reproducible.[5-7] However, these tests measure only the interactions between the organism and antibiotics. The tests do not assess the patient's underlying disease or immunologic response to infection, the pharmacokinetic properties of the antibiotics, the rapidity in which effective antimicrobial therapy is initiated, or the other numerous factors which clearly influence the eventual outcome of the patient-organism confrontation.[12]

Despite the limitations of *in vitro* susceptibility tests, the clinical data indicate a good correlation between the test results and patient response. This is illustrated by

Antimicrobial Susceptibility Testing, Edited by
J.A. Poupard *et al.*, Plenum Press, New York, 1994

examining the relationship between the *in vitro* tests and bacteriological response for patients treated with nitrofurantoin, cefotetan, cefotaxime, or ciprofloxacin (Tables 1-4). In each example, patients infected with susceptible organisms had a better bacteriological response compared with patients infected with resistant organisms. It should be noted, however, that despite the good agreement, more than half of the patients with "resistant" organisms responded to apparently inappropriate antibiotics. This observation illustrates the limitations of the *in vitro* diagnostic and therapeutic tests.

LABORATORY CONSIDERATIONS

The laboratory is also confronted with a number of difficult issues, not the least of which is when should susceptibility tests be performed and by what method? It is beyond the scope of this review to establish criteria for performing antimicrobial susceptibility tests. Rather I would emphasize the need to assess carefully, in collaboration with infectious disease specialists, the significance of organisms isolated in the clinical laboratory. Consensus for the need to test specific isolates and testing frequency should be developed by both medical as well as laboratory personnel. If this is done, then the number of tests performed and the costs associated with *in vitro* susceptibility testing can be appropriately controlled. The impact of continuous monitoring the appropriateness of *in vitro* susceptibility tests is illustrated by the experience in my laboratory during the last ten years. During this time period, the volume of blood cultures and urine cultures increased while antimicrobial susceptibility tests actually declined (Figure 1).

An important decision that confronts the laboratory is what testing method should be used. The two most common procedures currently used in the clinical laboratory are Kirby-Bauer disk diffusion and variations of broth microdilution. These methods are frequently referred to as qualitative and quantitative susceptibility tests, respectively. However, both tests are quantitative measurements of antibiotic activity. The difference between the tests is the format used to report the results. Antibiotic activity in the Kirby-Bauer test is determined by measuring continuous zones of inhibited growth around the antibiotic disks, but the results are reported qualitatively as susceptible, moderately susceptible or intermediate, or resistant. Although the microdilution results are commonly reported as minimum inhibitory concentrations (MICs), the results could also be reported qualitatively. Indeed, this would be appropriate when only one to three antibiotic concentrations are tested as is common in some commercially prepared susceptibility test panels.

The decision to use either the disk diffusion or the microdilution test method is complex. From 1982 to 1988 there was a dramatic increase in microdilution or MIC testing (Figure 2). Although the number of microdilution tests has stabilized since 1988, 70 to 75% of the US hospital laboratories are currently using some variation of the microdilution susceptibility test method (unpublished data, Dr. Robert Kreisler). Presumably this shift from disk diffusion to microdilution has been generated by the belief that microdilution tests are better predictors of clinical outcome, more reproducible, or more cost-effective. However, the clinical efficacy data presented above demonstrate that microdilution tests are no better than disk diffusion tests (e.g., data for cefotaxime) and possibly less accurate (e.g., ciprofloxacin data). Likewise, reproducibility studies with the disk diffusion and microdilution tests demonstrate better than 95% test reproducibility for both test methods (5-7). One of the major

TABLE 1. CORRELATION BETWEEN NITROFURANTOIN SUSCEPTIBILITY RESULTS AND IN VIVO RESPONSE

Susceptibility Results	Number of Patients	Percent with Bacteriologic Cure
≤32 ug/ml (susc)	41	90
64 ug/ml (intermed)	11	73
≥128 ug/ml (res)	30	50

Data submitted to NCCLS

TABLE 2. CORRELATION BETWEEN CEFOTETAN SUSCEPTIBILITY RESULTS AND IN VIVO RESPONSE

Susceptibility Results	Number of Patients	Percent with Bacteriologic Cure
Susceptible	2536	94
Moderately susceptible	82	96
Resistant	212	77

Data submitted to NCCLS

TABLE 3. CORRELATION BETWEEN CEFOTAXIME SUSCEPTIBILITY RESULTS AND IN VIVO RESPONSE

Susceptibility Results	Number of Patients	Percent with Bacteriologic Cure
≤8 ug/ml (susc)	1440	93
16-32 ug/ml (intermed)	99	78
≥64 ug/ml (res)	26	65
≥23 mm (susc)	1591	92
15-22 mm (intermed)	177	83
≤14 mm (res)	49	63

Data submitted to NCCLS

TABLE 4. CORRELATION BETWEEN CIPROFLOXACIN SUSCEPTIBILITY RESULTS AND IN VIVO RESPONSE

Susceptibility Results	Number of Patients	Percent with Bacteriologic Cure
≤1 ug/ml (susc)	722	89
2 ug/ml (intermed)	39	90
≥4 ug/ml (res)	14	86
≥21 mm (susc)	1815	91
16-20 mm (intermed)	115	79
≤15 mm (res)	13	62

Data submitted to NCCLS

TABLE 5. USAGE VOLUME AND COSTS OF IN VITRO SUSCEPTIBILITY TEST SYSTEMS

Susceptibility Test Product	Unit Volume x 10^6 (%)	Sales Volume ($) x 10^6 (%)
K-B disk diffusion	6.0 (30)	9.0 (14)
Baxter MicroScan (susc only)	5.0 (25)	20.0 (30)
Baxter MicroScan (ID/susc)	2.0 (10)	9.0 (14)
Vitek Systems	3.5 (17.5)	14.0 (21)
API UniScept	1.2 (6)	4.8 (7)
Other[a] (susc. only)	1.5 (7.5)	6.0 (9)
Other[a] (ID/susc)	0.8 (4)	3.6 (5)

(a) Other products include Micro-Media, Difco/Pasco, BBL Systems
(Data provided by Dr. Robert Kreisler)

TABLE 6. DETECTION OF AMINOGLYCOSIDE RESISTANT ENTEROCOCCI

Method	Aminoglycoside (conc; ug/ml)	Percent False-Susceptibility at: 24 h	48 h
Standard microdilution	Gentamicin (500)	3	3
	Gentamicin (2,000)	16	3
	Streptomycin (2,000)	14	14
Pasco microdilution	Gentamicin (2,000)	3	0
	Streptomycin (2,000)	6	3
Vitek	Gentamicin (500)	19	-
	Streptomycin (2,000)	0	-

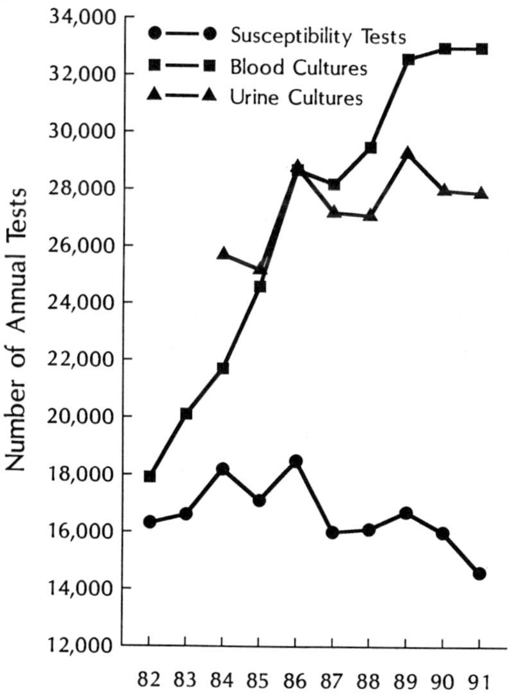

Figure 1. Number of Antimicrobial Susceptibility Tests, Blood Cultures, and Urine Cultures Performed Anually at Saint Louis from 1982 to 1991.

causes of testing problems is preparation of the test inoculum, which is performed in essentially the same manner for both test methods. The cost effectiveness of the tests is difficult to assess because instrument and consumable material costs, as well as personnel expenses, will vary widely. It was estimated that 20 million susceptibility tests were performed in 1990, with 30% by disk diffusion and the remainder by microdilution methods (Table 5). This represents a market of approximately $66,400,000. Despite a unit sales volume of 30%, the disk diffusion tests accounted for only $9,000,000 of sales or 14% of the market. This is attributed to the significantly higher costs associated with commercially prepared microdilution tests, which with volume discounting is approximately $4.00 per susceptibility test tray and $4.50 for the combination identification and susceptibility test tray. The cost of the inoculum tray and inoculator, as well as instrument depreciation and maintenance, will significantly increase the testing costs compared with a consumable cost of approximately $1.50 for disk diffusion testing. It is unlikely that this high cost of consumable supplies with dilution testing can be off-set by reduced personnel costs. These comments are not meant to disparage commercial susceptibility test systems, but rather to recognize that the convenience of prepared microdilution trays, automated interpretation of the test results, and computer integrated reporting systems comes at a price.[2]

As mentioned above, the correlation between *in vitro* test and *in vivo* response is not perfect. Although patient factors and antibiotic pharmacokinetic properties are responsible for many of these discrepancies, specific testing problems have also been identified. For example, poor susceptibility test reproducibility should be anticipated for results distributed near interpretive breakpoints (e.g. *Pseudomonas aeruginosa* with gentamicin; Figure 3). In this example, minor changes in the measured zone of inhibited growth can shift a susceptible organism into the intermediate or resistant categories. This problem can be further confounded by the selection of the interpretive breakpoints, as is illustrated by examining susceptibility tests with mezlocillin. In December, 1988 the NCCLS changed the interpretive standards for mezlocillin tests with *Enterobacteriaceae*: the zone of inhibited growth defining resistance was changed from $\leq$12 mm to $\leq$17 mm; susceptibility from $\geq$16 mm to $\geq$21 mm. The impact of this change in my hospital was that Enterobacteriaceae susceptible to mezlocillin decreased from 86% to 73%, even though the relative distribution of strains did not vary over this two year period (Figure 4). The test results were particularly dramatic with *Klebsiella* strains, because the majority of the organisms were distributed between 18 and 25 mm (Figure 5). Like the example cited above with *Pseudomonas* and gentamicin, minor changes in the test results with *Klebsiella* and mezlocillin had a major impact on the test interpretation. A similar change in the interpretive standards was introduced in 1988 for tests with mezlocillin and *Pseudomonas*: resistant changed from $\leq$12 mm to $\leq$14 mm and susceptible from $\geq$16 mm to $\geq$18 mm. The result of this apparently minor 2 mm change was a 10% decrease in strains classified as mezlocillin susceptible in my hospital (Figure 6). More recently the intermediate category was eliminated for tests with *Pseudomonas* and mezlocillin. With this change it is commonplace for organisms distributed near the 18 mm interpretive breakpoint to shift between susceptible and resistant.

Traditional testing methods may not be useful for certain organisms. A well-known example of this is *in vitro* testing of enterococci for the detection of aminoglycoside resistance.[4,10,11] It is recognized that susceptibility to a high concentration of gentamicin (500 to 2,000 ug/ml) or streptomycin (2,000 ug/ml) predicts susceptibility to the synergistic combination of the aminoglycoside with a cell-wall active antibiotic. Conventional disk diffusion tests cannot predict enterococcal

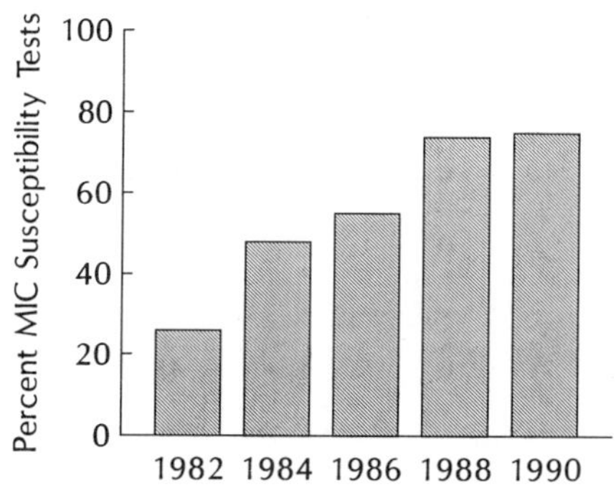

Figure 2. Percent of Quantitative Broth or Agar Dilution Tests Performed in the United States.

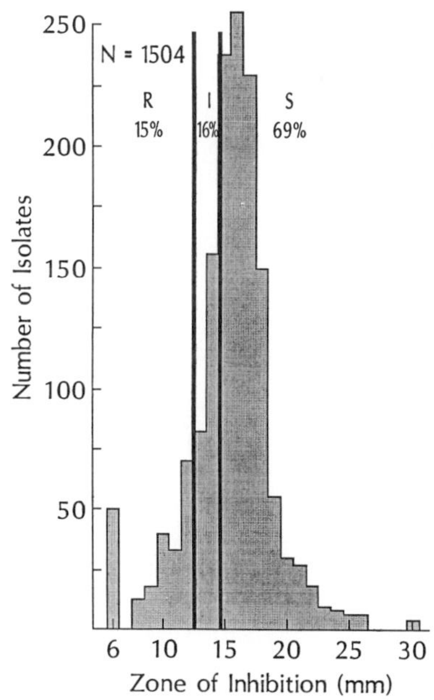

Figure 3. Analysis of Antimicrobial Susceptibility Tests with Pseudomonas aeruginosa Against Gentamicin Performed in 1991 at Barnes Hospital. The Peak Population of Organisms are Distributed Next to the Interpretative Breakpoint Separating Organisms Classified as Susceptible and Intermediate.

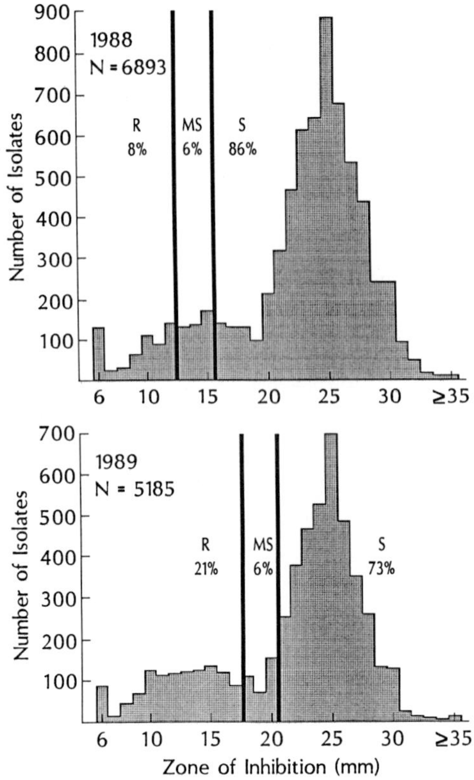

Figaure 4. Analysis of Antimicrobial Susceptibility Tests With All Enterbacteriaceae Isolates and Mezlocillin Performed in 1988 (top panel) and 1989 (bottom panel) at Barnes Hospital. The Change in Percent Susceptible or Resistant Isolates is Solely Due to the Change in Interpretive Standards.

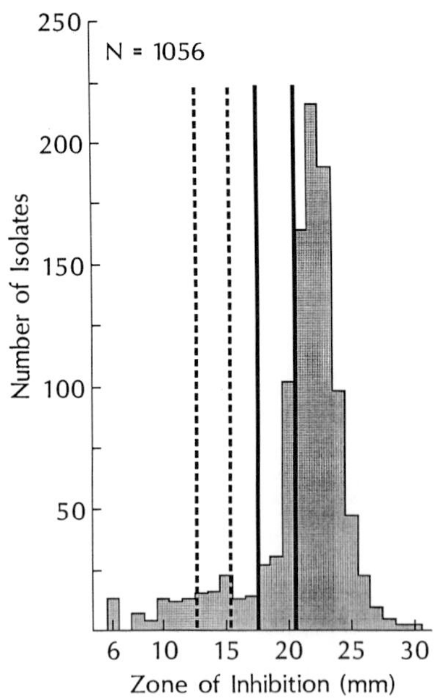

Figure 5. Analysis of Antimicrobial Susceptibility Tests with Klebsiella Against Mezlocillin Performed in 1989 at Barnes Hospital. By Shifting the Original Interpretive Standards (dashed lines) to Higher Zone Sizes (solid lines), the Percent of Klebsiella Susceptible to Mezlocillin Decreases Significiantly.

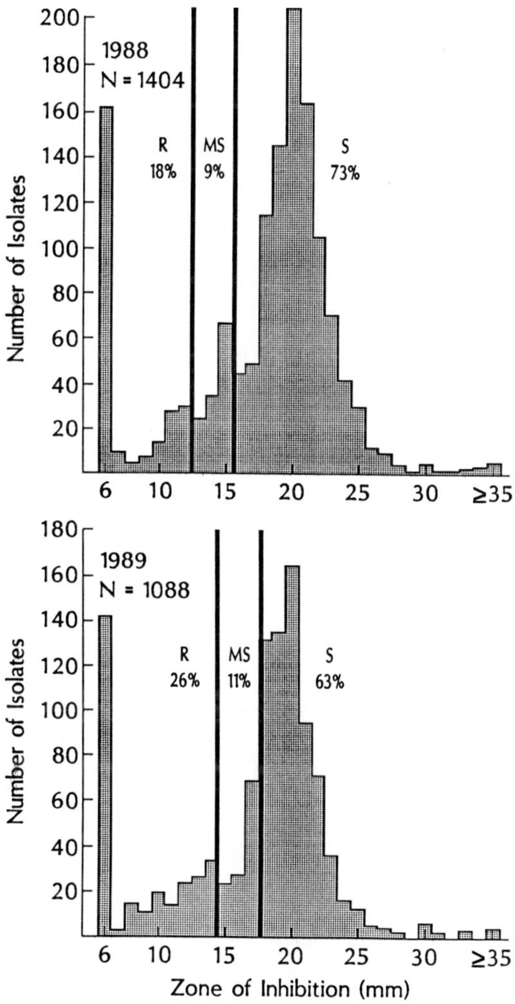

Figure 6. Analysis of Antimicrobial Susceptibility Tests with Pseudomonas aeruginosa Against Mezlocillin Performed in 1988 (top panel) and 1989 (bottom panel) at Barnes Hospital. As with Klebsiella-Mezlocillin Tests, the Change in Interpretive Standards Dramatically Reduces the Percent of "Susceptible" Strains.

susceptibility to these aminoglycoside concentrations.[10] Likewise, manual or automated microdilution tests are frequently unable to detect aminoglycoside resistance (Table 6), unless the testing conditions are carefully controlled.[4,11] Similar problems have also been reported for detecting oxacillin resistance in staphylococci. [1,3,13]

Finally, surveys of users of commercially prepared susceptibility test systems have identified additional problems. Whereas the laboratory using the disk diffusion test can select virtually any licensed antibiotic for *in vitro* testing, the users of commercially prepared microdilution trays are restricted to the antibiotics selected by the commercial supplier. Although custom prepared trays are available from the larger suppliers, these trays are more expensive.

Antimicrobial susceptibility testing of fastidious organisms can be problematic with both disk diffusion and microdilution tests. Although, commercial manufacturers have developed specialized test broths, alternative methods (e.g., macrotube broth dilution) must frequently be selected.

Automated susceptibility test systems may offer the convenience of a self-contained incubator and growth monitoring system, but this will also restrict the number of tests that can be processed by one instrument. In practice, this is generally not a problem with susceptibility test systems unless microbial identification and other functions are performed simultaneously by the instrument.

Complete automation of susceptibility testing from the preparation of the inoculum to the compilation of the results has not been developed. Until this is accomplished, significant personnel savings will not be realized.

Adequate quality control of microdilution panels has also been difficult. Quality control organisms must be selected with MIC values in the mid-range of the tested antibiotic concentrations. This is generally not possible for all antibiotics, so a panel of organisms must be selected. If only 1 to 3 concentrations of the antibiotics are tested (as is frequently the situation when tests with a large number of antibiotics are offered in one panel), then meaningful quality control testing is virtually impossible to achieve.

CONCLUSIONS

Both disk diffusion and microdilution tests are excellent predictors of clinical response to antimicrobial therapy - within the limitations of the tests. The selection of the specific testing method that is used in a laboratory must be individualized. Although automated systems offer a number of conveniences, including test standardization, rapidity, a computerized interfacing with other hospital reporting systems, this comes at an increase in costs. It is important to remember that whatever susceptibility test method a laboratory selects, it will have its limitations. Recognition of these limitations is vital for informed interpretation of the testing results.

REFERENCES

1. H.F. Chambers, Methicillin-Resistant Staphylococci. Clin Microbiol Rev 1:173-186 (1988).

2. C.E. Cherubin, R. Eng, M. Appleman, A Critique of Semi-automated Susceptibility Systems. Rev Infect Dis 9:655-659 (1987).
3. P.R. Coudon, D.L. Jones, H.P. Dalton, G.L. Archer, Evaluation of Laboratory Tests for Detection of Methicillin-Resistant Staphylococcus aureus and Staphylococcus epidermidis. J Clin Microbiol 24:764-769 (1986).
4. S.A. Fuller, D.E. Low, A.E. Simor, Evaluation of a Commercial Microtiter System (MicroScan) Using both Frozen and Freeze-Dried Panels for Detection of High Level Aminoglycoside Resistance in Enterococcus spp. J Clin Microbiol 28:1051-1053 (1990).
5. R.N. Jones, Antimicrobial Susceptibility Testing (AST): A Review of Changing Trends, Quality Control Guidelines, Test Accuracy, and Recommendation for the Testing of Beta-Lactam Drugs Diagn Microbiol Infect Dis 1:1-24 (1983).
6. R.N. Jones, D.C. Edson, and the CAP Microbiology Resource Committee, Interlaboratory Performance of Disk Agar Diffusion and Dilution Antimicrobial Susceptibility Tests, 1979-1981. Am J Clin Pathol 78:651-658, (1982).
7. P.R. Murray, J.R. Zeitinger, D.J. Krogstad, Reliability of Disk Diffusion Susceptibility Testing. Infect Control 3:230-237, (1982).
8. National Committee for Clinical Laboratory Standards. Performance Standards for Antimicrobial Disk Susceptibility Tests, 4th Ed., Approved Standard. NCCLS Document M2-A4. Villanova, PA: 1990.
9. National Committee for Clinical Laboratory Standards. Methods for Dilution Antimicrobial Susceptibility Tests for Bacteria that Grow Aerobically, 2nd Ed., Approved Standard. NCCLS Document M7-A2. Villanova, PA: 1990.
10. Pfaller MA, AC Niles, PR Murray. Evaluation of the Kirby-Bauer Disk Diffusion Test as a Screening Test for High Level Aminoglycoside Resistance in Enterococci. Am J Clin Pathol 82:458-460, (1984).
11. Sahm DF, S Boonlayangoor, JE Schulz. Detection of High-Level Aminoglycoside Resistance in Enterococci other than Enterococcus faecalis. J Clin Microbiol 29:2595-2598 (1991).
12. Washington JA. Discrepancies between In Vitro Activity of and In Vivo Response to Antimicrobial Agents, Diagn Microbiol Infect Dis 1:25-31 (1983).
13. Woolfrey BF, RT Lally, MN Ederer. Evaluation of the AutoMicrobic System for Detection of Resistance of Staphylococcus aureus to Methicillin. J Clin Microbiol 19:464-467 (1984).

CLINICIAN UTILIZATION OF RAPID ANTIBIOTIC SUSCEPTIBILITY DATA: A PROSPECTIVE STUDY

Franklin P. Koontz

University of Iowa Hospitals, Iowa City, Iowa

INTRODUCTION

The University of Iowa Hospitals is a 1,000 bed tertiary care center which serves as a regional hospital for hematology-oncology, cystic fibrosis and a variety of chronic disease patients. It also houses diverse transplantation patients, ranging from kidney, heart, liver-pancreas, to bone marrow. Thus a fair percentage of our patient load is compromised by immunosuppression or chronic disease which influences the demand for rapid testing and equally rapid reporting systems. In the areas of bacterial identification and susceptibility tests (MICs), we utilize the Vitek System as the mainstay of the laboratory due to its rapid turnaround time and reproducible accuracy. If we encounter isolates not in the Vitek database we use the manual API for idents and B-D Sceptor for MICs. I'll try to detail this as we follow a specimen through our work-flow pattern. To fully appreciate the Iowa System of rapid testing and reporting, one must understand that we are one of the most automated and instrumented laboratories in the country. All steps of specimen handling from admit or rejection, to final report, are computer mediated which gives us reproducibility as well as rapid, accurate test results. When the specimen is received in the lab and entered into our system via the lab computer, a work-flow card listing the media and stain, etc., required for that specific type of specimen is printed out. This will vary with specimen types, e.g., a wound would get routine culture and Gram stain only, however an abscess would also get anaerobic culture, while a BAL would demand all of these plus culture and stains for fungi or TB, if from the appropriate patient load. Thus, the depth to which a given specimen is examined is determined by the specimen type, source, patient status, presence of antibiotics, etc. Simultaneous to that "print-out", the specimen and all the tests ordered will appear on the CRTs in the clinical services area as "Results Pending." At the moment that any test result (Gram stain, Giemsa, Latex Antigen) is generated in the lab, it also appears on those clinical service CRTs. Thus, the clinician knows what we know at the moment we know it. The one exception to this is the result of automated MIC or ID from the Vitek which is interfaced to the hospital mainframe computer (IBM), so that data goes over to the IBM automatically without

Antimicrobial Susceptibility Testing, Edited by
J.A. Poupard *et al.*, Plenum Press, New York, 1994

the technologists having to manually enter the result as they did for the stains, latex tests, probes, etc. Of course the supervisory technologists are alerted to monitor those interfaced results from the Vitek which said supervisor can approve (finalize) or reject (amend), or add on additional data. So the data the clinician sees can change from Gram stain result, to preliminary ID, to final ID with MIC results in a matter of hours. The advantage here is obvious. We are producing results to the clinician in a time span short enough for our results to directly influence patient management in a variety of cost accountable ways: initiating more specific less broad coverage of antimicrobials, negating additional specimens to all labs while searching for a diagnosis or infectious agent, and resulting in the patient leaving an intensive care unit or the hospital sooner. This point is critical, the money savings will occur in the pharmacy or the nursing services, etc., the laboratory may not always see the savings, but the health care system overall will save money.

I should digress here to emphasize that the computer is programmed to accept a limit of specimens of certain types per day or per week. This results in a moderate rate of specimen rejection. We routinely will accept only one sputum or urine for culture per patient per day. If the patient has been manipulated or procedured, e.g., "induced sputum, thoracic aspirate, etc." for sputum, or had a catheterized UA, then that additional specimen would be accepted. We do not need to do multiple sputums and urines per patient, per day. I only accept three sputa per week, per patient, for TB or fungi. In the case of blood cultures I will accept 2-3 culture sets per day, the fourth set, etc., will be rejected. Theoretically, one should only accept 2-3 sets for 2-3 days, then one should wait two days to see what happens. If no positives, repeat that procedure then stop and re-evaluate. One should not run 2-3 sets per day ad nauseum. I have not as of today initiated a rejection policy on how many days "total" that I will accept blood cultures because I'm not sure of the cutoff or magic number. I'll be honest with you, the Bone Marrow Transplant Service is limited on the number of blood culture sets per day, but not on the number of days per week, per month, or per hospital stay. Those patients are so brittle, they could go septic in the "winking of an eye."

Therefore, I'm much more flexible in my rejection rules relative to the bone marrow transplant people. To increase the effect of this rapid automated system we utilize other instruments in tandem with the Vitek and the computer reporting system. Sterile body fluids except urine and blood, are routinely examined by using the Shandon Cytospin Centrifuge to prepare smears for Gram, Giemsa or special stains (FA, etc.). The increased direct detection rate for pathogens by cytospin versus standard conical tube centrifugation ranged from 20-28% for pleural fluid and CSF respectively. The cytospin centrifuges directly onto the slide with a filter paper interface that absorbs all the soluble proteins, etc., producing a smear-stain with a totally clear background which makes the causative organism or WBCs jump out at you visually. We also improved our "buffy coat" results by cytospin of the buffy coat from a standard centrifuged tube of blood.

For blood and ascitic fluid cultures (spontaneous peritonitis) we tandem the BACTEC NR 660 to the Vitek. Positive BACTEC bottles are entered via syringe and a ten ml aliquot is centrifuged, Gram stained, re-suspended in saline and put directly into the appropriate Vitek card(s). If the organism is a Gram-negative rod both the ID and MIC cards are inoculated, however, if Gram-positive, only an MIC card is used, as the direct ID on Gram-positive cocci has some obvious problems. The cards are marked for the catalase or oxidase reaction on the basis of the Gram stain morphology.

We do not attempt direct coagulase or thermonuclease tests since the inoculum is usually too dilute to give a valid result. The result of the Gram stain morphology is entered into the computer and the submitting service also receives a telephone report of that result. (We automatically phone results of all positive blood cultures and CSF Gram stains as well as computer reporting.) For the BACTEC, we routinely use the "BACTEC Plus 26-27" system, the high volume 10 ml blood inoculum bottle with resins. This system increased our detection of septicemia by 35% overall (all patients), 53% if the patient was on antibiotics at the time of blood draw, and 19% if the patient had not received antimicrobial within the previous 10 days. Thus, increased yield via the BACTEC means increased rapid ID and MIC via the Vitek. The two instruments work synergistically, so to speak. I should emphasize that we also streak out the positive BACTEC bottles in case of polymicrobic sepsis and to have the organism for further testing (cidal levels, synergy tests, etc.). On average, fewer than 5% of the blood isolates examined in this way require repeating the Vitek analyses. Thus, the direct BACTEC to Vitek System saves at least one day in getting ID-MIC results 95% of the time. In the 5% re-tested, no time is really lost as we use the pure subculture as we previously did years ago. So the direct inoculation of Vitek Cards from the BACTEC is really not a gamble at all.

There is also obvious time gain when one works with isolated colonies from other specimens (sputa, urine, wounds, etc.). Dual cards (ID and MIC) are inoculated from the same suspension and the results obtained in 4-10 hours via the Vitek, and immediately reported via the Vitek-IBM computer interface to the clinician in the hospital or clinics. Usually the Gram-positives report out in 4 hours, 6-8 hours for enterics, and 7-10 hours for pseudomonas. Some isolates only get an MIC via Vitek as we still ID quite a few of the GNR by "scratch and sniff." I mean a large beta hemolytic colony that smells like mothballs or a wet woolen suit gets a touch indol test. If positive the result E. coli is entered and no Vitek ID is required. The same type of quickie manual ID is used quite often for the *Proteus*, *Pseudomonas aeruginosa*, and *Staphylococcus aureus*.

I should emphasize here that we routinely run the GPI card on coagulase negative staphylococci from blood, CSF, line sites, etc. This is not done on sputa, wounds, etc. The reasoning here is that some of the *S. epidermidis* group (*S. epidermidis*, *S. warneri*, *S. haemolyticus*) are now the number one cause of nosocomial sepsis in the U.S. and the species ID of the coagulase negative staph (CNS) helps the clinician in therapy decisions.

We have had 6 isolates of *S. haemolyticus* that were truly vancomycin resistant. The ID also helps to determine valid isolates from contaminants on blood cultures, Hickman lines or line sites, etc., as the three species mentioned above have been proven pathogenic in those sites while all the other CNS (cohnii, capitis, etc.) have not. The only questionable isolate at this time is *S. hominis* as it has been shown to be the causative bug in a few instances but is usually a skin contaminant. We do *S. saprophyticus* identifications on urines for epidemiologic reasons.

Table 1 shows the rate of increase of the CNS over the past four decades at the University of Iowa Hospitals. The reasons for the rapid increase in the last decade are the indwelling lines and monitors. As you can easily see, the greatest jump has been during the "bone marrow transplant" era. The rest of the nosocomial sepsis top five are *S. aureus*, *enterococci*, *Candida sp.* and *E. coli* in that order. This is a marked change from a few years ago when *E. coli*, K-E Group, *Pseudomonas*, Proteus Group,

Table 1. Infection Rate for Nosocomial Bloodstream Inductions Due to Coagulase Negative Staph (CNS).

University of Iowa Hospitals

Year	Rate*	Percent of Nosocomial Sepsis
1945-55	0.6	<1%
1974-75	1.3	<1%
1980	5.2	8%
1984	14.6	10%
1987	42.0	26%

* Rate - Cases per 10,000 admissions.

Martin et al. - *Ann Intern Med.* 110:9-16, 1989.

and *S. aureus* were the top five. The reasons for the changes are the increased use of indwelling lines, third generation cephalosporins, and immunosuppressive agents. So actually the organisms didn't change, the condition of the host changed.

Okay, that's the general idea of how we do our automated test procedures and computer reporting, but how does all this impact on the clinicians usage of antibiotics? When our MIC data is brought up on the IBM-CRT screen by the clinician, they also see a daily cost code, e.g., ampicillin $2.00, cefazolin $22.00, cefoxitin $43.00 and ceftazidime S114.00, etc., per day cost to the Pharmacy, and whether that agent is on "restricted antibiotic list" which means the clinician must fill out a special prescription form. If the clinician probes "dose chart" on the screen with the light pen, the pharmaco-kinetic data for all antibiotics is shown on the screen, based upon a healthy adult of 150 pounds with adequate renal function. The data suggests the usual dosages by all three routes, if applicable, and the levels of the drug usually found in various body compartments, e.g., urine, blood, CSF based upon the suggested doses at the recommended dosing intervals. Thus, we utilize the most accurate rapid methods coupled with rapid computerized reporting supplemented with additional patient care guidance data. Obviously then we do not have the usual antibiotic usage pattern problems seen in other hospitals . . . or do we?

We designed a covert daily chart review of four in-service areas to document the clinician's response to the lab data. This program used the Infections Control nurses and Quality Assurance nurses as they performed daily service rounds. We trained them specifically on what to look for in the patients' charts to determine from the progress notes, 1A1 Form, lab hard copy reports, etc., what the clinician did with the data. It should be emphasized that this method is the only way one can covertly document physician utilization of lab data. It is impossible, however, to always know whether the clinicians were unaware of the data available to them or they saw it and chose to ignore it. Almost 1,200 patients were reviewed daily for almost three months (6,500 patient days) and specialists from the Infectious Diseases Division reviewed the culture MICs and which antibiotics were actually used for therapy on that patient. Appropriateness was also evaluated on the basis of loci of the infection, toxicity, half-life of the drug and cost.

The results of the study were less than a fabulous illustration of the cooperative efforts of the health care system. The lab end looked great as the average time for preliminary data to appear on the computer was 18-19 hours after specimen collection. Final reports, including species identification and the MIC values, averaged 22-31 hours post specimen collection, however we could only document that the clinician saw the MIC about 10% of the time in the 24 hours following the posting of that data on the computer and in the patient's chart. For documentation we accepted a comment in the progress notes, a change in antibiotics reflecting the MIC data, comments or "doodles" on the chart lab copy, and in one instance a coffee cup ring on the chart copy (if they put their coffee on it, they must have seen it!!). So much for the age old adage, "I don't care what its name is, what kills it?" Faced with this data, we then had the specialists from Infectious Disease review the patients' charts, retrospectively and in detail for the appropriateness of the therapy given. Table 10 shows the results of this primary review of the first 723 antibiotic decisions (each antibiotic is a decision, e.g., if the patient was on three antibiotics and the MIC listed those three plus six others, that equals nine decisions). The table speaks for itself. The most common error (disagree) was, for example, the patient was receiving vancomycin, gentamycin, clindamycin for empiric therapy, we isolated *E. coli* from the wound and the clinician continued the vancomycin and clindamycin. That would equal two (2) disagreements. If the clinician discontinued the vancomycin, clindamycin, and gentamycin and started cefazolin or continued the gentamycin only, then that equaled agreement. If he/she discontinued the three drugs and put the patient on ceftazidime or imipenem then that is a judgment call.

Table 2a. Appropriate Therapy*.

	G. Surg	Med	Peds	Urol	Total
Agree	32	154	126	76	388
Disagree**	3	64	85	24	176
Judgement***	12	52	71	24	159

* 723 decisions analyzed by infectious disease physician.

** Disgaree - wrong drug, drug not needed.

*** Judgement - drug OK, but better or cheaper should be used.

This data disturbed us so much that we did a similar detailed review of an additional 507 decisions on patients where knowledge of the MIC report had been documented. These results are presented in the following table. Please note that percentage of agreement actually declined. (I guess this means that a clinician is more prone to make an error on antibiotic choices if he/she knows more data on the organism or its MIC).

Also note that on these two tables, surgeons have the best compliance rate. Now this was shocking as half of the surgeons I've ever met don't believe in the "germ theory of disease."

Their high rate of compliance was actually due to the fact that *S. aureus* was their major pathogen and nafcilln was selected empirically and continued after the MIC report was posted. Conversely, pediatrics and medicine had many more Gram-negative organisms requiring a more detailed decision making process.

Table 2b. Appropriate Use of Antibiotics* after MIC Report.

(507 Decisions Evaluated)

	G. Surg	Med	Peds	Urol	Total
Agree Rx	24	80	74	46	224 (43%)
Disagree Rx	13	55	77	21	166 (33%)
Judgement**	6	37	60	14	117 (24%)
Total	43	172	211	78	507

* Patients on 1-5 drugs, not including prophylaxis.

** Judgement - less costly or better choice could have been made.

Table 3. Documented Cultures with Inappropriate Response Frequency M.D. Notes Results and Makes at Least One Inappropriate Decision.

G. Surg + Urol	Med	Peds	Total
8/24	29/39	30/47	61/110*
(33%)	(74%)	(64%)	(56%)

55 positive cultures, 55 negative cultures = 110

Finally, we waited a month, then reviewed the antibiotic therapy decisions in the same patient areas on 110 patients randomly, but evenly, split between "culture-positive" and "culture-negative" result groups. Again, the surgeons did the best for the same reason already discussed and the inappropriate therapy as previously mentioned was most often related to leaving the patient on antibiotics that were not justified even though the appropriate antibiotic was also being used. The reason for this occurrence is simply that the patient was improving on "multiple drug therapy" and the clinician was not about to "fix what ain't broke." There were a few instances when totally inappropriate therapy was given, e.g., ceftazidime for a *S. aureus* where the MIC was "off the wall."

It is relatively simple to ascertain that the cost of health care is markedly influenced by this indiscriminate usage pattern for certain antibiotics, but there is also a greater problem here. During the time of the study and subsequent to it, we have verified an unbelievable increase in resistance to the antibiotics most commonly left in the therapy regimen but not justified or judgmentally a bad decision. The following table shows what has happened to ceftazidime over the past four to five years in this hospital relative to ceftazidime usage and abusage. Note that in 1987 we stopped reporting MIC results on that drug except for Pseudomonas and enterics resistant to cefotaxime and ceftriaxone, etc. The abusage and selected resistance continued. In 1988-89 we got it under control with "strict restrictions" on its usage, but that policy only lasted about one year and then we started the process all over again, except for *Pseudomonas* which has remained susceptible at the 90% level. Could this mean they

Table 4. Problems with Ceftazidime - UIHC Doses Used Versus Percent Susceptible.

	1986	1987*	1988	1989**	1990***	1991
Doses Used	5,135	23,000	26,000	18,500	27,000	26,000
Citrobacter sp.	95%	75%	60%	95%	77%	67%
Enterobacter sp.	95%	72%	52%	73%	76%	62%
Ps. aeruginosa	95%	85%	72%	88%	88%	88%

* Selective reporting - No Effect.

** Stringent usage policy enforced - Reversed Trend.

*** Policy on usage relaxed - Resistancy Reappears (FPK 1991)

are using that drug for everything except *Pseudomonas* which is the one bug it should be used against. Only Dan Quayle knows for sure.

Over the past 2-3 years we have also noted an increase in the MICs of the coagulase negative *staphylococci* to vancomycin. We have had six patients with vancomycin resistant (MIC >32) *S. haemolyticus*. All were chronic renal failure, on dialysis, receiving prophylactic vancomycin weekly. I have no problem with these instances, they are classic examples of God and Darwin working in tandem. What concerns me is the pattern of vancomycin against this species in other patients where the average MIC is rising at an alarming rate. The following data illustrates my

Table 5. S. *haemolyticus* MIC Values (%) 1989-1991.

MIC	1989	1991
<1 μg	37%	9%
1-2 μg	56%	63%
3-4 μg	7%	28%
>4 μg	0%	0%

concerns. We did not have any resistant strains during the past two years, but note that the percent susceptible at breakpoint (4 μg) is four-fold higher than two years ago. Are we heading for disaster?

Now what do we as clinical microbiologists do about these problems in utilization of our hard earned data? We must get more involved in the continuing education of the clinicians by participating in hospital "in-service programs," increase our "one-one" interface with clinicians, get involved in hospital committees that relate to antibiotic usage and abusage. We need to educate the clinician on the nuances of antibiotic usage. Most clinicians were not taught this material in medical school. They were for cardiovascular drugs, etc., but not antibiotics. To solve this problem the laboratory and the ID physician are going to have to train and/or retrain a large proportion of the medical staff if our efforts are truly going to impact upon patient care management in this cost containment, "dammit Darwin" era of health care delivery.

WHEN WE SHOULD BE TESTING, HOW OFTEN AND WHAT TO REPORT

Raymond C. Bartlett

Hartford Hospital
Hartford, Connecticut

INTRODUCTION

This chapter will offer practical suggestions for clinical microbiologists to help establish cost-effective policies for antimicrobial susceptibility testing of bacteria and efficient reporting of information to guide antimicrobial therapy. The cost of antimicrobic therapy greatly exceeds the cost of the bacteriology laboratory. Therefore, the combined efforts of the microbiologist, the pharmacist and the medical staff should be focused on producing information that reduces antimicrobic cost.[1] Procedures will be recommended here to minimize the production of clinically unnecessary information and the reporting of excessive data that is potentially confusing to clinicians. The practices described are based on over 30 years of cooperative effort between the Microbiology Laboratory, Pharmacy Services and the Division of Infectious Diseases, Department of Medicine at Hartford Hospital. These efforts have reduced the percentage of antimicrobic cost in our pharmacy budget to about 15%. Nationally it is more typical for 30-50% of pharmacy budgets to be consumed by antimicrobic use.[2,3] Further, about 20,000,000 bacterial isolates are subject to antimicrobial susceptibility testing in the nation's laboratories annually.[4] Compared to one other major center of comparable size and volume of testing we perform one fifth the number of antimicrobial susceptibility tests (Table 1). This results largely from controls that we have imposed on the quality of specimens and the extent of testing that is performed.[5] Studies have shown that the cost for microbiologic processing in our laboratory is lower than that for conventional processing without controls.[6]

Table 1. Antimicrobial Susceptibility Testing at Hartford Hospital.

Specimens/month				6,617
Susc. Tests Performed	1 Isolate	2 Isolates	3 Isolates	Total
	241	36	4	281
Tests not performed < 5 days since previous submission				22
				303

Antimicrobial Susceptibility Testing, Edited by
J.A. Poupard *et al.*, Plenum Press, New York, 1994

Physicians must understand the need for continuous interaction with the laboratory in the submission and processing of microbiologic specimens. If the physician does not pay attention to preliminary laboratory reports, important messages may be missed and further processing may be delayed or terminated using the systems recommended here. Physicians on the staffs of hospitals where the laboratory does not employ these systems easily become accustomed to submitting specimens and then waiting to look up the report four or five days later if patients are not doing well. This assumes that all specimens are carried through an equally complete processing protocol. A cost-effective approach often requires that there be interaction between the laboratory and the clinician at various steps from the initial direct examination through preliminary assessment of the culture and selection of isolates for complete identification and susceptibility testing. Ten or twenty years ago it was appropriate for physicians to assume that specimens would be given the maximum extent of examination that technology would allow and that the reported information could be reviewed by the physician for clinically relevant information. This approach is no longer economically feasible. Either some limit must be placed on the extent to the processing of all specimens, or there must be some interaction to establish the clinical usefulness of applying progressive effort and expense to the examination of specimens. The future will bring laboratory information systems that will employ physician interactive ordering and interpretation of results. Today physicians would view interaction with a laboratory information system as a nuisance. They are used to having clerks take care of filling out request slips. Eventually they will obtain so much useful information to assist diagnosis and treatment that the initiative to install such a system may actually come from them rather than the laboratory. Until such systems become fully tested and efficient, the approaches described in this chapter will provide the most that can be accomplished with current laboratory information processing.

WHEN TO TEST: INAPPROPRIATE SPECIMENS

It is important that antimicrobial susceptibility testing be applied only to isolates from specimens containing material that is associated with infection. Bacteria representing normal flora or colonization should not be tested. Examples include isolation of *Hemophilus influenzae*, *Staphylococcus aureus* or *Streptococcus pneumoniae* from throat swabs. Although these organisms may cause otitis media, sinusitis and lower respiratory tract infections, they are a part of the indigenous flora of the throat and should not be tested for susceptibility to guide the treatment of infections. Swabs collected from the external nares or the nasopharynx also are not capable of yielding information that will predict the etiology of sinusitis or otitis media. Exceptions include studies of colonization which are sometimes conducted among groups of immunocompromised patients to identify predominating colonizers to enable prompt treatment of infection developing in such patients. This practice should be carefully coordinated with the clinicians responsible for the patients to assure that inappropriate treatment does not occur. Another situation is in the evaluation of an outbreak of cross infection when it is important to identify sources of colonizing bacteria which may be implicated. Such studies should always be supervised and approved by the Infection Control Committee.

Similarly, isolates obtained from specimens collected from superficial skin lesions such as decubiti, colostomy and tracheostomy stomata, the mouth and genital mucous membranes such as the vagina or the urethral meatus generally yield indigenous and colonizing bacteria that are of questionable relationship to any underlying infection. Paradoxically, the amount of effort involved in isolating and testing isolates from such specimens is vastly larger and more costly than what is expended on specimens containing bacteria that can lead to a report that will properly influence therapy. Some laboratories report identification and susceptibilities of only one or two of the predominating isolates from cultures of such specimens. This is especially misleading and likely to misdirect antimicrobial therapy.

If the laboratory has been in the practice of processing such specimens, a review should be conducted with appropriate representatives of the medical staff. A policy should be agreed upon to preserve the specimen at 40°C for up to a week and render a report to the requesting physician requesting that he/she contact the director of the laboratory to review the indications for processing. Prior to implementation of such a policy, staff physicians should be fully informed of the rationale through written newsletters and conferences. An objection that may be voiced is the deterioration and delay that might occur while awaiting such a consultation. Physicians need to appreciate that it would be extremely unusual for processing of such a specimen to provide useful information. Such an infrequency would not outweigh the cost and would have a potentially confusing effect on reporting identification and susceptibility information from these kinds of specimens. Transport containers making use of semisolid media always should be used to provide better preservation of all swab collected specimens.

It is necessary to know what clinical condition physicians are concerned with when they submit swab specimens from the vagina or uterine cervix. The laboratory request slip should provide boxes labeled "vaginitis", "gonorrhea", "surgical wound" or endometritis". Otherwise the technologist does not know how to culture the specimen or which isolates should be considered pathogens. It should be agreed upon with the medical staff that specimens will be held pending consultation from the physician if no indication of the clinical condition is given.

Other types of specimens may represent material from the site of infection, but the infections are predictably mixed and identification and susceptibility testing does not alter therapy. Examples include perirectal and pilonidal abscesses and fistulae, and Bartholin's abscess. All are treated by drainage. Dozens of species may be isolated if cultures are conducted. If antibiotics are to be used, they should be directed as intestinal bacteria and not be chosen on the basis of the bacteriology report.

MICROSCOPIC EVALUATION OF SPECIMENS

All swab-collected specimens of respiratory or wound exudate should be evaluated for evidence of acute inflammatory cells through preparation of a Gram stained direct smear. A number of schemata for evaluating sputum specimens have been described based on the presence of squamous cells.[5] Squamous cells are found only above the level of the vocal cords. An abundance indicates that the specimen has become contaminated with these cells and that mouth and throat bacterial flora have been introduced. Such specimens should not be processed. The report should indicate evidence of oropharyngeal contamination and request recollection or a consultation from the physician if unanticipated clinical criteria justify processing. We have applied these criteria also to wound and cervicovaginal exudate where culture is performed despite an abundance of squamous cells and a paucity or absence of neutrophils because the isolation of such organisms as *Streptococcus pyogenes* or *Clostridium perfringens* may assist in the rapid recognition of serious life threatening infections. Specimens of exudate that contain an abundance of squamous cells may yield enteric bacteria, *Staphylococcus spp.* and nonfermentative gram negative rods of dubious relationship to infection. In such cases, the general nature of the isolates may be reported

```
11/29 0500 WOUND - EXUDATE ABDOMEN
11/29         DIRECT SMEAR - Q3
                CELLS SEEN                      BACTERIA IN SMEAR
                NEUTROPHILS - MANY              ENTERICS
                SQUAM - NONE                    BACTEROIDES-HEMOPHILUS
                                                STREP
                                                CLOSTRIDIUM
              FINAL REPORT
                GRAM STAINED DIRECT SMEAR SUGGESTS THE PRESENCE OF A
                MIXTURE OF ORGANISMS OF INTESTINAL ORIGIN.  A MIXED
                CULTURE CAN BE ANTICIPATED.  PLEASE CONSULT MICROBIOLOGY
                IF CULTURE OF THIS SPECIMEN CAN BE EXPECTED TO PRODUCE
                USEFUL INFORMATION.
```

Figure 1. Report of exudate displaying multiple morphologic types of bacteria in direct smear. Culture adds cost and little additional useful infromation.

based on gross morphology of the colonies with a request for consultation for complete identification and susceptibility testing if clinically indicated.

Specimens Displaying Mixtures of Bacteria in Smear

Specimens from sites contaminated with intestinal bacteria often display an abundance of bacterial forms in the gram stained direct smear. Typical examples are appendiceal abscesses, and draining fistulae. Other examples are superficial wounds which may be colonized or contaminated with either or both cutaneous and intestinal flora. If more than three discretely different types are seen, both gram positive and gram negative, and at least one morphologically suggesting anaerobic bacteria, there is little value in culture. A report should be rendered of the findings suggestive of a mixed infection originating from intestinal bacteria (Figure 1). This will be a more rapid and less costly means of directing appropriate therapy than spending several days and much laboratory effort on the isolation and susceptibility testing of only that portion of the microbial population that can be isolated in pure culture. Here again, culture and reporting of only the predominating isolate is misleading. The finding of end to end pairs of slender evenly staining gram negative rods suggestive of *Pseudomonas* should be reported and these specimens should be cultured because of the resistance that this organism usually displays and the impact that its presence often has on selection of therapy.

MIXED CULTURES

Specimens that appear to be uncontaminated in the gram stained smear may yield a large number of isolates that suggest a mixed infection. In general, we believe that isolation of more than three pathogens from such specimens should result in a consultation for identification and susceptibility testing unless correlation with the bacteria observed in the gram stained direct smear can help identify any of the isolates as more likely to be clinically significant than the other (Figure 2). Note that we report the morphologic types of bacteria seen in the smear in terms that are more understandable to physicians than the usual bacteriologic terms "gram positive cocci", "gram negative rods", etc. It is expensive and

```
11/29 0500 WOUND - EXUDATE ABDOMEN
11/30        DIRECT SMEAR - Q3
               CELLS SEEN                  BACTERIA IN SMEAR
               NEUTROPHILS - MANY          ENTERICS
               SQUAM - NONE                BACTEROIDES-HEMOPHILUS
                                           STREP
                                           STAPH
             CULTURE - FINAL REPORT
               SMEAR RESULTS (ABOVE) AND CULTURE RESULTS (BELOW)
               INDICATE MULTIPLE POSSIBLE PATHOGENS.  EITHER EMPIRIC
               THERAPY SHOULD BE USED OR ANOTHER SPECIMEN SHOULD BE
               COLLECTED BY A PROCEDURE THAT AVOIDS COLONIZATION.
               PLEASE CONSULT MICROBIOLOGY FOR FURTHER PROCESSING.
               E. COLI
               PROTEUS SP.
               PSEUDOMONAS SP
               KLEBSIELLA/ENTEROBACTER SP
```

Figure 2. This good quality specimen yields four types of bacteria in both direct smear and culture. Although all are potential pathogens, the specimen should not be used to guide diagnosis and treatment as stated.

time-consuming to test large numbers of isolates, and the approach to treatment should be empirical in such cases. As increasing numbers of pathogens are isolated, the probability increases that a more diligent search would yield even larger numbers. This emphasizes the wisdom of empirical antimicrobial therapy using broad spectrum agents for the categories of organisms isolated. Cumitech 2 suggests that clean catch urine specimens containing three or more isolates may be contaminated.[7] We studied this problem at Hartford Hospital and found that in only 10% of instances did either clean catch or closed catheter urine specimens display the same mixture of isolates when another specimen was collected within 24 hours.[8] It has become our practice to report the gross colony morphology without identification and susceptibility testing when more than one species is isolated 105 CFU/ML (Figure 3). When two different colony types of gram negative rods are seen they should be identified because often they turn out to be variants of the same species. If they are different species, the report suggests that the specimen may be contaminated and requests recollection or consultation for processing. If a second specimen contains the same mixture, the isolates should be identified because they are much more likely to be clinically significant. In other published studies conducted here these approaches to the handling of urine specimens have been shown to contribute more than any other reductions in cost in our laboratory.[6]

```
11/29 0500 URINE - CLEAN CATCH
11/30
             CULTURE - FINAL REPORT
               SERRATIA MARCESCENS 100,000/ML
               STAPH SP. 100,000/ML
               THIS IS A MIXED CULTURE SUGGESTING THE PROBABILITY OF
               CONTAMINATION.  PLEASE RESUBMIT OR CONSULT MICROBIOLOGY
               FOR PROCESSING.
```

Figure 3. Urine specimens containing two or more species exceeding 10^5 CFU/ml are most likely contaminated and should be recollected.

SPECIMENS FROM NORMALLY STERILE BODY SITES

Almost any isolate from a normally sterile body site should be considered a potential pathogen. This problem has intensified in recent years with the increasing association with infection of organisms such as *Corynebacterium* and coagulase negative *staphylococci*. Maximum effort should be made to handle such specimens under laminar flow hoods using careful technique to minimize contamination. This is especially important in laboratories in institutions where there are many immunocompromised patients or patients who have been fitted with prosthetic and intravascular devices. Frequently, the isolation of one colony of an organism from only one of several plates will suggest laboratory contamination. In most cases, it is probably wise to report this and request a consultation for further identification and susceptibility testing if clinically indicated. It is frequently helpful to call the physician to discuss the probable significance of the isolate.

WHAT ISOLATES TO TEST

Having dealt with appropriate specimens and the relevance to infection of the organisms isolated, the next question is which isolates should be tested. Only species that are unpredictable in susceptibility and that yield reproducible and accurate results with conventional methods should be tested. There are some common exceptions which require modifications in procedure that will be described below. In general, however, there is no need for routine testing of *Streptococcus spp.* except *Enterococcus sp.*, *Streptococcus pyogenes* from throat cultures or *Streptococcus pneumoniae* from sputum specimens do not require routine testing. Very infrequently, testing may be required of erythromycin or other alternative agents that are not uniformly effective against these two species when patients are allergic to penicillin. The increasing frequency of *S. pneumoniae* isolates relatively resistant to penicillin has led most microbiologists to test isolates from blood and spinal fluid for susceptibility to 0.06 mcg/ml. This is most reliably performed using Mueller-Hinton broth with 5% lysed horse blood.[9] Growth at this concentration indicates that the MIC be determined to classify the level of insusceptibility. One may be selective in testing the susceptibility of coagulase negative *staphylococci*. They should be tested when isolated from normally sterile sites in pure culture such as spinal fluid or blood. Consultations should be requested for many other situations in which this organism is isolated. When isolated from intravascular catheter tips, susceptibles should not be reported because it encourages unnecessary antimicrobial therapy, often with vancomycin. Catheter tip sepsis with this organism is often effectively treated by removal of the catheter. Susceptibility testing also is not indicated when this species is isolated from mixed cultures where its pathogenicity is questionable.

Cornybacterium spp. should be considered significant pathogens when they are isolated from normally sterile sites. *C. jeikeium* and the D2 strain are highly resistant. Susceptibility testing is often helpful in supporting the identification of one of these strains. It is best to call the clinician to review the probable clinical significance of such an isolate and the need for testing specific antimicrobics such as vancomycin, rifampin and ciprofloxacin.

There are many infrequently isolated pathogenic species that should not be routinely tested for a variety of reasons. These are listed in Table 2. Either their susceptibility is predictable and/or results of *in vitro* testing are unreliable. In some instances testing may be dangerous and increase the risk of laboratory acquired infections such as with *Brucella*, *Francisella*, and *Pseudomonas pseudomallei*. It is best to contact clinicians when these

Table 2. Species with Predictable Antimicrobic Susceptibility: Testing is Not Required.

Aeromonas sp.	Moraxella catarrhalis
Brucella*	Campylobacter
Capnocytophaga*	Cardiobacterium*
Eikenella*	Gardnerella*
Francisella tularensis*	Legionella *
Listeria monocytogenes*	Ps. pseudomallei
Pasteurella multocida*	Yersinia enterocolitica
Yersinia pestis	Y. pseudotuberculosis
Vibrio parahemolyticus*	

* also may be inaccurate

organisms are isolated to review with them the appropriate therapy. Otherwise, they are likely to call the laboratory, often after regular working hours, inquiring why susceptibility testing was not performed and complaining that the lack of results is compromising treatment.

There should be no need to test routinely *Neiserria meningitidis* or *N. gonorrhoeae* in most laboratories, although routine testing of the latter species for beta lactamase is customary. Public health laboratories may wish to screen susceptibility for epidemiologic reasons. Neither treatment or prophylaxis of *N. meningitidis* depends on *in vitro* test results. Ceftriaxone is uniformly effective for *N. gonorrhoeae*, and penicillin continues to be effective for *N. meningitidis* although there are extremely rare reported cases of penicillin resistance.[10] There is a need to screen isolates for the dissemination of such a strain, but this should be the role of public health laboratories.

WHAT ANTIMICROBICS TO TEST

The selection of antimicrobial agents to test can be a daunting task. The table provided by NCCLS[11] is of some help but still provides a large range of options which, if followed, is likely to result in testing of more agents, and the reporting of more results, than is necessary. Laboratories should offer routine susceptibility testing batteries for four different groups of isolates: 1) *Staphylococcus sp.*, 2) *Enterococcus sp.*, 3) Enteric bacteria, and 4) *Pseudomonas aeruginosa*. These tests may be performed by the method of Bauer et al., or one of the many automated and semiautomated commercial systems, or by microdilution. Those agents tested in our laboratory are listed in Table 3. Many of the agents are tested only because they are included in the reagent cards provided by the manufacturer. It is increasingly important that agents be tested that can be given orally to help reduce the unnecessary cost of parenteral antimicrobial therapy. When results are reported for urinary tract isolates, only results should be reported for drugs which are applicable to the treatment of urinary tract infection. Special panels of agents may be tested following specific consultation with knowledgeable clinicians where the unique activity of high levels of concentration of tetracyclines, for example, may be effective in the treatment of *Pseudomonas* urinary tract infection. Agents such as nitrofurantoin and indanyl carbenicillin that are only effective when used to treat urinary tract infection should be reported only for urinary tract isolates. Indications for testing the newest antimicrobial agents will vary widely from one institution to another. The choice for testing should be closely coordinated with the therapeutics committee and representatives of clinical services

Table 3. Antimicrobics Tested: Hartford Hospital Routine.

Routine

Staphylococcus sp.	
Cephalothin	Oxacillin
Ciprofloxacin	Penicillin
Clindamycin	Tetracycline *
Cotrimoxazole +	Vancomycin
Erythromycin	(beta lactamase)
Ampicillin-sulbactam *	
Enterococcus sp	
Ampicillin	Nitrofurantoin
Cephalothin *	Norfloxacin
Chloramphenicol	Penicillin
Clindamycin	Streptomycin
Erythromycin	Tetracycline *
Gentamicin	Vancomycin
Gram negative	
Ampicillin	Gentamicin
Ampicillin-sulbactam	Ciprofloxacin
Cefazolin	Cotrimoxazole
Cefotaxime	Imipenem
Cefotetan	Tobramycin
Ceftazidime	
Pseudomonas	
Amikacin	Imipenem
Aztreonam	Mezlocillin
Ceftazidime	Piperacillin
Ciprofloxacin	Ticarcillin
Gentamicin	Tobramycin

Selective: urine

Ampicillin	Ciprofloxacin	Tetracycline
Amikacin	Cotrimoxazole	Tobramycin
Augmentin	Gentamicin	
Carbenicillin	Mezlocillin	
Cefoxitin	Nitrofurantoin	
Ceftriaxone	Norfloxacin	
Cephalothin		

* Tested only because included in test card by manufacturer. See tables 11, 12 and 13 for results to be reported.

+ Trimethoprim-sulfamethoxazole

knowledgeable in management of infectious diseases. Testing, and reporting, as will be discussed later, should be coordinated with the formulary, and frequent routine testing and reporting of susceptibility of drugs that are not available without special consultation should be avoided. The agents that frequently fall into these categories are listed in Table 4.

HOW OFTEN TO TEST

For many years it has been our practice not to repeat antimicrobial susceptibility testing on the same isolate from the same body site if reisolated from a series of consecutively submitted specimens within five days. A message is issued making reference to the most recently reported susceptibility data and requesting consultation for testing if needed (Figure 4). Isolates are held for seven days pending such a request. An informal poll of other laboratories indicates that testing every other day is a frequent option when daily submissions occur. The growing concern over the appearance of stably derepressed Type 1 beta lactamase producing gram negative rods has suggested that more frequent testing of such isolates may be indicated. A recent experience in our hospital showed how quickly acquisition of this type of resistance can occur (Figure 5). Currently we test *Acinetobacter* sp., *Pseudomonas aeruginosa*, and *Enterobacteriaceae*, except *E. coli*, whenever they are reisolated from daily consecutively submitted specimens. Continued testing of gram positive bacteria no more often than once every five days when they are repeatedly reisolated seems warranted because acquisition of resistance is not as frequent or as rapid.

Table 4. Antimicrobics for which Use is Frequently Restricted.

Aztreonam
Imipenem
Ampicillin/sulbactam (Unasyn)
Ceftazidime
Ticarcillin/clavulanic acid (Timentin)
Vancomycin
Quinolones
Azithromycin

```
11/29 0715 WOUND - EXUDATE ABDOMEN
11/30         DIRECT SMEAR - Q3
                CELLS SEEN                       BACTERIA IN SMEAR
                NEUTROPHILS - MANY               STAPH
                SQUAM - NONE
              CULTURE - FINAL REPORT
                STAPH AUREUS
                SEE ANTIMICROBIAL SUSCEPTIBILITY PREVIOUSLY REPORTED OR
                CONSULT MICROBIOLOGY FOR TESTING
```

Figure 4. When *S. aureus* is isolated repeatedly from specimens resubmitted within 5 days testing is not repeated.

Respiratory secretion specimens submitted

Date	Isolate	AMP	AUG	CFZ	CIP	COT	GEN	IMI	TAN	TAX	TAZ	TOB	UNA	Therapy
		Susceptibilities												
9/19	Acinet	R		R	S	S	S	S	R	S	S	S	S	CFZ, GEN
9/20	Acinet	R		R	S	S	S	S	I	S	S	S	S	CFZ ↓, GEN ↓
"	Klebs	S		S	S	S	S	S	I	S	S	S	S	
"	H infl	S	S		S	I				S	S		S	
9/22	Acinet	NO SIGNIFICANT CHANGE FROM PREVIOUS SPECIMEN												CFZ ↓, GEN ↓, TAX
"	Klebs	SEE SUSCEPTIBILITIES REPORTED 9/20												
"	H infl													
9/27	Acinet	I		R	S	S	S	S	R	S	S	S	S	CFZ ↓, GEN ↓, TAX ↓
"	Citrob	R		R	S	R	R	S	R	I	R	R	R	
"	Enterob	R		R	S	R	R	S	R	R	R	R	R	
9/29	Acinet	NO SIGNIFICANT CHANGE FROM PREVIOUS SPECIMEN												CFZ ↓, GEN ↓, TAX ↓
"	Citrob	SEE SUSCEPTIBILITIES REPORTED 9/27												
"	Enterob													CIP
10/7	Acinet	I		R	R	R	R	R	R	I	R	R	R	CIP ↓
"	Enterob	R		R	S	R	R	R	R	R	R	R	R	

Figure 5. M.L.T. age 43 admission diagnosis, head trauma Complication, pneumonia.

Policy in effect not to repeat susceptibility testing of gram negative rods repeatedly isolated delayed recognition of acquisition of resistance by *Acinetobacter calcoaceticus var anitratus*

AMP=ampicillin; AUG=augmentin, CFZ=cefazolin; CIP=ciprofloxacin; COT=cotrimoxazole; GEN=gentamicin; IMI=imipenem; TAN=cefotetan; TAX=cefotaxime; TAZ=ceftazidime; TOB=tobramycin; UNA=unasyn

Acinet=*Acinetobacter*; Klebs=*Klebsiella pneumoniae*; H.infl.=*Hemophilus influenzae*; Citrob=*Citrobacter sp*; Enterob=*Enterobacter cloacae*

HOW TO REPORT RESULTS

Most microbiologists report susceptibility as "S", "I", or "R". Most clinicians are familiar with this because of its routine application to the Kirby-Bauer test. When MIC testing was first introduced, it was rapidly learned that MIC reports could not be interpreted by clinicians without provision of elaborate interpretative data. With time it has been learned that MICs are best reported as S I R through translation using NCCLS recommended breakpoints.[11] Clinicians frequently will be confused by the NCCLS designation "moderately susceptible". We avoid the use of the term and further reference will be made to this practice below. Some microbiologists have employed the system described by Ellner employing the "inhibitory quotient".[12] The probable mean peak blood level of the drug was made the numerator, and the MIC the denominator, to derive a quotient that would represent the magnitude of inhibitory effect that could be anticipated. Although clever and clinically sophisticated, the approach has not gained wide acceptance. Some have recommended that the daily cost for both the drug and its administration be included on reports (Figure 6). Microbiologists should continue to experiment with such reporting and the readability of the format by obtaining frequent feedback from the pharmacy service and representatives of the medical staff. Some computer reporting

MICROBIOLOGY LABORATORY RESULTS
GENERAL HOSPITAL

(Biographic data omitted)

Organism #1 Staphylococcus aureus (STA)

Organism #2 Haemophilus influenzae (HIN)

Antibiotics	STA(1)	HIN(2)	Adult dosage	Achievable serum level mcg/ml	Hospital Cost Code 1=lowest 10=highest
Ampicillin		R	PO 500 mg IV 500 mg	2.5 - 4.0 2.9 - 7.1	1 4
Ampicillin/ sulbactam	≤4 S		IV 3000 mg Am 1000 mg sulbac	109 - 150	4
Cefoxitin		R	IV 1000 mg	110	10

(remainder of report omitted)

Figure 6. Example of a report provided by an anonymous laboratory. It lists results of antimicrobial susceptibility testing including MIC data. "SIR" breakpoint designations, adult doasage, achievable levels and relative cost 6 antimicrobics.

systems generate relatively cumbersome displays of susceptibility data. Although users may become gradually familiar with their appearance, a maximum effort should be made to render reports as compact and readable as possible. Two different alternative formats for reporting data in our laboratory are displayed in Figures 7 and 8. The abbreviations appear on the reverse side of the printed report. When multiple isolates are found and their susceptibility is reported, the readability of the format becomes increasingly important.

WHAT TO REPORT?

Which results to report?

A substantial battery of antimicrobics may be used for testing, but this does not mean that all results should be reported. Frequently the species identification of isolates is not known when susceptibility testing is set up so it is difficult to be selective about which antimicrobics to test. The report should encourage the use of the most clinically appropriate antimicrobics and discourage the use of those that are least appropriate. It is essential that the microbiologist, representatives of the pharmacy service and the medical staff, and a physician subspecializing in infectious disease participate jointly in the process of deciding which antimicrobic test results to report. This will be influenced by the antimicrobics that are in the institution's formulary and ones for which use is restricted. Technologists working in small hospital laboratories sometimes complain that no such group can be brought together in their hospital and that reporting of antimicrobics is uncontrolled and irrational. Usage may be dictated primarily by the interest of pharmaceutical company

```
11/29 0500 RESP. - SEC - LEUKENS
12/03      DIRECT SMEAR - Q3
             CELLS SEEN                            BACTERIA IN SMEAR
             NEUTROPHILS - MANY                    PSEUDOMONAS
             SQUAM - NONE                          BACTEROIDES-HEMOPHILUS
           CULTURE - FINAL REPORT                  STAPH
             PSEUDOMONAS AERUGINOSA
               AMP - R, AUG - R, AZT - S, CFZ - R, CIP - S, CLI - R,
               COT - R, GEN - S, IMI - S, NAF - R, PIP - S, TAX - R,
               TAZ - S, TOB - S, UNA - R
             HEMOPHILUS INFLUENZAE
             BETA LACTAMASE NEGATIVE
               AMP - S, AUG - S, AZT - S, CFZ - R, CIP - I, CLI - R,
               COT - I, GEN - I, IMI - S, NAF - R, PIP - S, TAX - S,
               TAZ - S, TOB - I, UNA - R
             STAPHYLOCOCCUS AUREUS
               AMP - R, AUG - S, AZT - R, CFZ - S, CIP - I, CLI - S,
               COT - S, GEN - I, IMI - S, NAF - S, PIP - R, TAX - S,
               TAZ - R, TOB - I, UNA - S
             S-I-R MAY BE BASED ON LOCAL EXPERIENCE OR
             PUBLISHED DATA.  PLEASE CONSULT MICROBIOLOGY
             FOR TESTING
```

Figure 7. Hartford Hospital report of mixed culture of *S. aureus, Pseudomonas* and *H. influenzae*. All antimicrobics reported for each isolate are also reported for the others. Note Imipenem reported because of collective resistance of the three isolates to the other antimicrobics. It is difficult to read this report to readily detect the antimicrobics that are effective against all isolates.

```
11/29 0500 RESP. - SEC - LEUKENS
12/03    DIRECT SMEAR - Q3
           CELLS SEEN                            BACTERIA IN SMEAR
           NEUTROPHILS - MANY                    PSEUDOMONAS
           SQUAM - NONE                          BACTEROIDES-HEMOPHILUS
         CULTURE - FINAL REPORT                  STAPH
           1-PSEUDOMONAS AERUGINOSA
           2-HEMOPHILUS INFLUENZAE
             BETA LACTAMASE NEGATIVE
           3-STAPHYLOCOCCUS AUREUS
           S-I-R MAY BE BASED ON LOCAL EXPERIENCE OR
           PUBLISHED DATA.  PLEASE CONSULT MICROBIOLOGY
           FOR TESTING
```

	AUG	AMP	CFZ	AZT	CIP	CLI	COT	GEN	IMI	NAF	PIP	TAX	TAZ	TOB	UNA
1	R	R	R	S	S	R	R	S	S	R	S	R	S	S	R
2	S	S	R	S	S	R	I	I	S	R	S	S	S	I	S
3	S	R	S	R	I	S	S	I	S	S	R	S	R	I	S

Figure 8. Alternative format to report displayed in Figure 7.

<u>Key to Abbreviations in Figures 7 and 8</u>

AMP=ampicillin; AUG=augmentin; AZT=aztreonam; CFZ=cefazolin; CIP=ciprofloxacin; CLI=clindamycin; COT=cotrimoxazole (trimethoprim-sulfa); GEN=gentamicin; IMI=Imipenem; NAF=nafcillin; PIP=piperacillin; TAX=cefotaxime; TAZ=ceftazidime; TOB=tobramycin; UNA=unasyn representatives and the individual interests of the most influential members of the medical staff. The Joint Commission on Accreditation of Healthcare Organizations (JCAHO) is placing great pressure on all hospitals to regulate antimicrobic use. It would be surprising that even small hospitals have not felt this pressure. With or without the pressure of JCAHO, persons interested in improving antimicrobic use in hospitals should consider inviting speakers to address the issue before the medical staff. A bibliography of related publications would be helpful. Consulting groups are available to work conjointly with the medical staff to advise the pharmacy service and the laboratory in developing rational antimicrobic use practices.

Generic Terminology

Generic terminology should be used for antimicrobial agents as much as possible. This has become more difficult with the advent of such tongue twisters as ceftazidime, ceftizoxime, cefotaxime, and ceftriazone where the corresponding trade names Fortaz, Cefizox, Claforan, and Rocephin are much more easily remembered and distinguished from one another. We have had great difficulty reporting test results for trimethoprim /sulfamethoxazole. The trade names Bactrim and Septra are widely recognized by clinicians, but the generic term cotrimoxazole is not known to many. The abbreviation TMP/SMX is preferred by some. In the final analysis it is best that whatever reporting terminology is to be used be coordinated through consultation with the pharmacy service and the medical staff. If this leads to the decision to use trade names, it should be recognized that competitive bidding frequently will result in change in product and that the laboratory report may no longer reflect the actual commercial agent being used.

Reporting Results

If abbreviations are used, it is important that the report provide a key. Microbiologists should avoid reporting potentially misleading information. Some automated susceptibility testing systems have been observed to determine false susceptibility of *Klebsiella pneumoniae* to ampicillin. This should not be a problem if the laboratory never reports the results of susceptibility of this species to that drug. Some may argue that reporting of resistance of this species to ampicillin is important to discourage inexperienced house staff from using the drug empirically. If such a decision is made, then the result should be routinely reported as resistant regardless of actual test results.

Microbiologists should be sufficiently familiar with the antimicrobial susceptibility profile of most isolates to enable recognition of probable errors in identification or susceptibility testing when unexpected susceptibility results are obtained. For many years we employed a systematic quality control procedure to detect such errors, and its application is described elsewhere in this book. It is commonly recognized that isolates of *Staphylococcus aureus* should be reported resistant to all cephalosporins when oxacillin resistance is observed regardless of test results. There is some disagreement about the same approach to reporting of results of testing coagulase negative staphylococci. It is important for a microbiologist who is familiar with the mechanisms of action of most antimicrobial agents to review reports to detect other types of misleading information.

For example, reporting of susceptibility of an isolate of an oxacillin-resistant coagulase negative staphylococci to ampicillin/sulbactam could have occurred in our laboratory after a clinician had specifically requested that it be tested. We placed a qualifying statement in the report (Figure 9). A better solution would have been to have explained to the physician when the request was received that the antimicrobic would not have been appropriate to treat an infection with this organism regardless of *in vitro* test results.

Conditional Reporting

Conditional reporting also has been referred to as "cascade" reporting. This policy is widely practiced, and its application to various classes of drugs is illustrated in Table 8. For example, gentamicin is the primary aminoglycoside reported by most laboratories. Tobramycin susceptibility is reported only if the isolate is resistant to gentamicin. In turn, amikacin susceptibility is reported only if both gentamicin and tobramycin show resistance. Most computer reporting systems and automated antimicrobial test instruments provide the software to enable such reporting. Unfortunately, little attention has been given to application of such reporting to mixed cultures. The result may create a report which will cause clinicians to use combinations of antimicrobics to treat a mixed infection (Figure 10). In our laboratory this has led to reporting the test results of all antimicrobics for all isolates in a mixed culture that would be reported for any of the individual isolates (Figure 11). Computerized reporting systems and computerized software support for automated instruments do not enable such reporting, and this is a serious existing deficit.

```
08/26 1140 WOUND - ABDOMEN
08/29          DIRECT SMEAR - Q3
                 CELLS SEEN                     BACTERIA IN SMEAR
                 NEUTROPHILS - MANY             PSEUDOMONAS
                 SQUAM - NONE                   STAPH
               CULTURE - FINAL REPORT
                 PSEUDOMONAS AERUGINOSA
                   MEZ - S, CIP - S , GEN - S, TAZ - S, TOB - S
                 STAPH SPECIES NOT AUREUS
                   UNA - 100%, CFZ - 76%, NAF - 66%, CLI - 84%
                 PERCENT SUSCEPTIBILITIES ARE BASED ON
                 PUBLISHED DATA AND NOT ON TESTING OF THIS ISOLATE.
                 PLEASE CONSULT MICROBIOLOGY FOR TESTING

                 MIC RESULTS (MCG/ML) ON STAPH SPECIES NOT AUREUS
                 CFZ:  2    OXA:  16    UNA:  8
                 UNASYN PROBABLY SHOULD NOT BE USED TO TREAT STAPH
                 RESISTANT TO OXACILLIN BECAUSE MECHANISM IS NOT BETA
                 LACTAMASE (RCB)
```

Figure 9. Physician requested testing susceptibility of caogulase negative staphylococcus to unasyn (ampicillin-sulbactam). Although result suggests susceptibility, antimicrobic should not be used against oxacillin resistant strain (MIC = 16). Messages should be added to reports when results may mislead.

Table 8. Conditional Reporting: Only One Antimicrobic Usually is Reported Within Each Class.

Class

Cephalosporins: Cephalothin, cefazolin, cefotetan, cefotaxime, ceftazidime.
Aminoglycosides: Gentamicin, tobramycin, amikacin, (netilmicin)
Extend spectrum penicillins: mezlocillin, carbenicillin, ticarcillin, piperacillin, azlocillin

```
11/29 1140 RESP - SEC - LEUKENS
12/03         DIRECT SMEAR - Q3
                CELLS SEEN                    BACTERIA IN SMEAR
                NEUTROPHILS - MANY            MIXED MORPHOLOGY
                SQUAM - NONE
              CULTURE - FINAL REPORT
                PSEUDOMONAS AERUGINOSA
                  GEN - R, TOB - S, TAZ - S, PIP - I,
                CITROBACTER FREUNDII
                  GEN - S, TAX - S, PIP - S, COT - R
                MORGANELLA MORGANII
                  GEN - S, TAN - S, PIP - I, COT - R
                UPPER RESPIRATORY FLORA
```

Figure 10 Conditional reporting of results for individual isolates may prevent physician from recognizing any single antimicrobic that could be used to treat infection caused by all three isolates. Report encourages combination therapy. See key to Table 11 for abbreviations.

```
11/29 1140 RESP - SEC - LEUKENS
12/03         DIRECT SMEAR - Q3
                CELLS SEEN                    BACTERIA IN SMEAR
                NEUTROPHILS - MANY            MIXED MORPHOLOGY
                SQUAM - NONE
              CULTURE - FINAL REPORT
                PSEUDOMONAS AERUGINOSA
                  CIP - S, COT - R, GEN - R, PIP - I, TAN - R,
                  TAX - R, TAZ - S, TOB - S, UNA - R
                CITROBACTER FREUNDII
                  CIP - S, COT - S, GEN - S, PIP - S, TAN - R,
                  TAX - S, TAZ - S, TOB - S, UNA - R
                MORGANELLA MORGANII
                  CIP - S, COT - S, GEN - S, PIP - S, TAN - R,
                  TAX - S, TAZ - S, TOB - S, UNA - R
                UPPER RESPIRATORY FLORA
```

Figure 11 Improved version of report in Figure 10. All antimicrobics reported for any isolate are reported for others. Several choices of single antimicrobic therapy become apparent (Ciprofloxacin, Ceftazidime, and Tobramycin). Ciprofloxacin was not available when version in Figure 10 was reported. See Key to Table 9 for abbreviations.

REPORTING GUIDELINES FOR PERSONNEL

Laboratories equipped with laboratory information systems or instrumentation with software support have the capacity to program the reporting of test results for selected isolates. As mentioned, this will not enable customizing of reports for mixed cultures. Technologists can learn to work with tables that guide reporting for common isolates and frequent mixture of isolates. Those in use in our laboratory are shown in Tables 9 and 10.

Table 9. Guide to Reporting of Antimicrobic Susceptibility Test Results for Pure Cultures.

Sp.	Antimicrobic																	
	AMP	AUG	AZT	CFZ	CIP	CLI	COT	CUR	GEN	MET	NAF	PIP	TAN	TAX	TAZ	TOB	UNA	VAN
ESC	( )	---	---	( )	( )	---	( )	---	( )	---	---	---	( )	( )	( )	( )	---	---
KLE	---	---	---	((	( )	---	( )	---	( )	---	---	---	( )	( )	( )	( )	---	---
PMI	( )	---	---	( )	( )	---	( )	---	( )	---	---	---	( )	( )	( )	( )	---	---
ENB	---	---	---	( )	( )	---	( )	---	( )	---	---	---	---	( )	( )	( )	---	---
PSA	---	---	( )	---	( )	---	---	---	( )	---	---	( )	---	---	( )	( )	---	---
ACI	---	---	---	---	( )	---	( )	---	( )	---	---	---	---	( )	( )	( )	---	---
SER	---	---	---	---	( )	---	( )	---	( )	---	---	---	( )	( )	( )	( )	---	---
MOR	---	---	---	---	( )	---	( )	---	( )	---	---	---	( )	( )	( )	( )	---	---
PVU	---	---	---	---	( )	---	( )	---	( )	---	---	---	( )	( )	( )	( )	---	---
CIT	---	---	---	---	( )	---	( )	---	( )	---	---	---	---	( )	( )	( )	---	---
PRO	---	---	---	---	( )	---	( )	---	( )	---	---	---	---	( )	( )	( )	---	---
HEM	(R)	(S)	---	---	(S)	---	(I)	(S)	---	---	---	---	---	(S)	---	---	(S)	---
HEM	(S)	---	---	---	---	---	---	(S)	---	---	---	---	---	---	---	---	---	---
BFR	---	---	---	---	---	(I)	---	---	---	(S)	---	---	(S)	---	---	---	(S)	---
BFG	---	---	---	---	---	(I)	---	---	---	(S)	---	---	(I)	---	---	---	(S)	---
BSP	---	---	---	---	---	(S)	---	---	---	(S)	---	---	(S)	---	---	---	(S)	---
FNU	---	---	---	---	---	(S)	---	---	---	(S)	---	---	(S)	---	---	---	(S)	---
FSP	---	---	---	---	---	(I)	---	---	---	(S)	---	---	(S)	---	---	---	(I)	---
STA	---	---	---	---	---	( )	---	---	---	---	( )	---	---	---	---	---	---	---
STA	---	---	---	(R)	---	( )	---	---	---	---	(R)	---	---	---	---	---	---	( )
STN	---	---	---	( )	---	( )	---	---	---	---	( )	---	---	---	---	---	---	---
STN	---	---	---	( )	---	( )	---	---	---	---	(R)	---	---	---	---	---	---	( )

Table used in laboratory for guiding technologists in reporting results of susceptibility testing. Results of test are reported where brackets are shown. Test results are reported when bracket is empty. S, I, or R is reported without testing if displayed in bracket. See text for instructions regarding conditional reporting.

Species: ESC = *E. Coli*; KLE = *Klebsiella*; PMI = *P. mirabilis*
ENB = *Enterobacter*; PSA = *Ps. aeruginosa*; ACI = *Acinetobacter*; SER = *Serratia*
MOR= *Morganella*; PVU = *P. vulgaris*; CIT = *Citrobacter*; PRO = *Providencia*
HEM = *Hemophilus*; BFR = *Bacteroides fragilis*; BFG (*B. frag*. group) = *B. thetaiotaomicron*, *B. distasonis*, *B. vulgatus*, *B. ovatus*, *B. uniformis*; FNU = *Fusobacterium nucleatum*; FSP = *Fusobacterium species*; STA = *S. aureus*; STN = *S*. not *aureus*
Antimicrobic:AMP = ampicillin; AUG = augmentin; AZT = aztreonam; CFZ = cefazolin; CIP = ciprofloxacin; CLI = clindamycin; COT = cotrimoxazole (trimethoprim-sulfa); CUR = cefuroxime; GEN = gentamicin; MET = metronidazole
NAF = nafcillin; PIP = piperacillin; TAN = cefotetan; TAX = cefotaxime
TAZ = ceftazidime; TIM = timentin; TOB = tobramycin; UNA = unasyn; VAN = vancomycin

Table 10. Reporting Enterococcus Susceptibility (Pure Culture Only).

AMP MIC	<8 mcg/ml				>8 mcg/ml*
	urine	blood		other	all sources
GEN synergy		+	-		
REPORT					
AMP	S#	S#	I	S#	R
GEN		S#	R	S#	
NIT	X	0	0	0	X**
VAN			X		X
Add note **		+		+	

* = test for beta lactamase. Review with microbiologist drugs to report. X = report test result. ** = urine only
\# = NCCLS designation Moderately Susceptible
** = THERAPY SHOULD INCLUDE A PENICILLIN PLUS AN AMINOGLYCOSIDE

Reporting *Enterococcus sp.* test results has been made more complicated by the use of the designated "Moderately Susceptible (MS)" by NCCLS.[11] This was done to reduce the risk that physicians would treat serious systemic enterococcal infections with oral disease of ampicillin. Whether or not the reporting of MS would prevent such therapy is debatable. We believe it is more important to emphasize the need to use an aminoglycoside in combination with a penicillin or vancomycin. This recommendation is guided by testing for high level resistance to aminoglycosides. We restrict this to blood isolated and assume in our reporting that isolates from other sites can be treated as though they were susceptible to high levels of aminoglycosides (Figure 12).

When uncommon isolates or infrequently seen mixed cultures occur, the microbiologist should determine which results to report. It may be desirable to have a more complex algorithm to guide the microbiologist in this process especially if agreement has been reached about reporting in such situations with the pharmacy service and representatives of the medical staff. Otherwise, the microbiologist could be put in the position of having to call representatives of these groups each time an unusual report is being assessed or risk using his own judgment. Each institution will have to determine which is the best way to proceed. The algorithm agreed upon in our institution is complex and intended for use only by the microbiologist in charge (Table 11).

```
11/29 0500 BLOOD
12/03          GRAM STAIN
                 BACTERIA SUGGESTED - STREP
               CULTURE - FINAL REPORT
                 ENTEROCOCCUS
                   AMP - S, GEN - S
               THERAPY SHOULD INCLUDE A PENICILLIN PLUS AN AMINOGLYCOSIDE
```

Figure 12 Report of ampicillin susceptible *Enterococcus sp* from blood displaying susceptibility to a high level of gentamicin. NCCLS recommended designation moderately susceptible (MS) is not used. Instead use of aminoglycoside is recommended.

Table 11. Guide for Reporting of Susceptibility Results for Mixed Cultures.

Sp.	Antimicrobic																		
	AMP	AUG	AZT	CFZ	CIP	CLI	COT	CUR	GEN	IMI	MET	NAF	PIP	TAN	TAX	TAZ	TOB	UNA	VAN
ESC	( )	( )	( )	( )	( )	(R)	( )	( )	( )	( )	(R)	(R)	( )	( )	( )	( )	( )	( )	(R)
KLE	(R)	( )	( )	( )	( )	(R)	( )	( )	( )	( )	(R)	(R)	( )	( )	( )	( )	( )	( )	(R)
PMI	( )	( )	( )	( )	( )	(R)	( )	( )	( )	( )	(R)	(R)	( )	( )	( )	( )	( )	( )	(R)
ENB	(R)	( )	( )	( )	( )	(R)	( )	( )	( )	( )	(R)	(R)	( )	( )	( )	( )	( )	( )	(R)
PSA	(R)	( )	( )	( )	( )	(R)	(R)	(R)	( )	( )	(R)	(R)	( )	(R)	(R)	( )	( )	(R)	(R)
ACI	(R)	(R)	( )	(R)	( )	(R)	( )	( )	( )	( )	(R)	(R)	( )	( )	( )	( )	( )	( )	(R)
SER	(R)	( )	( )	(R)	( )	(R)	( )	( )	( )	( )	(R)	(R)	( )	( )	( )	( )	( )	( )	(R)
MOR	(R)	( )	( )	(R)	( )	(R)	( )	( )	( )	( )	(R)	(R)	( )	( )	( )	( )	( )	( )	(R)
PVU	(R)	( )	( )	(R)	( )	(R)	( )	( )	( )	( )	(R)	(R)	( )	( )	( )	( )	( )	( )	(R)
CIT	(R)	( )	( )	(R)	( )	(R)	( )	( )	( )	( )	(R)	(R)	( )	( )	( )	( )	( )	( )	(R)
PRO	(R)	( )	( )	(R)	( )	(R)	( )	( )	( )	( )	(R)	(R)	( )	( )	( )	( )	( )	( )	(R)
HEM	(R)	(S)	(S)	(R)	(S)	(R)	(I)	(S)	(I)	(S)	(R)	(R)	(R)	(R)	(S)	(S)	(I)	(S)	(R)
HEM	(S)	0	(S)	(R)	(S)	(R)	(I)	(S)	(I)	(S)	(R)	(R)	(R)	(R)	(S)	(S)	(I)	( )	(R)
BFR	(R)	(R)	(R)	(R)	(R)	(I)	(R]	(R)	(R)	(S)	(S)	(R)	(I)	(S)	(I)	(I)	(R)	(S)	(R)
BFG	(R)	(R)	(R)	(R)	(R)	(I)	(R]	(R)	(R)	(S)	(S)	(R)	(I)	(I)	(I)	(I)	(R)	(S)	(R)
BSP	(R)	(R)	(R)	(R)	(R)	(S)	(R]	(R)	(R)	(S)	(S)	(R)	(I)	(S)	(I)	(I)	(R)	(S)	(R)
FNU	(R)	(R)	(R)	(R)	(R)	(S)	(R)	(R)	(R)	(S)	(S)	(R)	(I)	(S)	(I)	(I)	(R)	(S)	(R)
FSP	(R)	(R)	(R)	(R)	(R)	(I)	(R)	(R)	(R)	(S)	(S)	(R)	(I)	(S)	(I)	(I)	(R)	(I)	(R)
STA	(R)	(S)	(R)	(S)	(R)	( )	( )	( )	(I)	(S)	(R)	( )	(R)	( )	(I)	(R)	(I)	(S)	0
STA	(R)	(R)	(R)	(R)	(I)	( )	( )	(R)	(I)	(R)	(R)	(R)	(R)	(R)	(R)	(R)	(I)	(R)	( )
STN	(R)	(R)	(R)	( )	(I)	( )	( )	( )	(I)	(S)	(R)	( )	(R)	( )	(I)	(R)	(I)	(S)	0
STN	(R)	(R)	(R)	( )	(I)	( )	( )	(R)	(I)	(R)	(R)	(R)	(R)	(R)	(R)	(R)	(I)	(R)	( )

Table used by Microbiologist to guide reporting of susceptibility for mixed cultures.(see figures 7-8 for example) S, I, or R is reported for all antimicrobics reported for any of the species isolated. Result obtained by testing is reported for empty bracket. S, I, or R result in bracket is reported without testing.

Species: ESC = E. coli; KLE = Klebsiella; PMI = P. mirabilis
ENB = Enterobacter; PSA = Ps. aeruginosa; ACI = Acinetobacter; SER = Serratia
MOR= Morganella; PVU = P. vulgaris; CIT = Citrobacter; PRO = Providencia
HEM = Hemophilus; BFR = Bacteroides fragilis; BFG = B. frg group = B. thetaiotaomicron, B. distasonis, B. vulgatus, B. ovatus, B. uniformis; FNU = Fusobacterium nucleatum; FSP = F. species; STA = S. aureus; STN = S. not aureus;
Antimicrobic:AMP = ampicillin; AUG = augmentin; AZT = aztreonam; CFZ = cefazolin; CIP = ciprofloxacin; CLI = clindamycin; COT = cotrimoxazole (trimethoprim-sulfa); CUR = cefuroxime; GEN = gentamicin; IMI = imipenem; MET = metronidazole; NAF = nafcillin; PIP = piperacillin; TAN = cefotetan; TAX = cefotaxime; TAZ = ceftazidime; TIM = timentin; TOB = tobramycin; UNA = unasyn; VAN = vancomycin

REPORTING RESTRICTED TO IN VITRO TEST RESULTS ?

Clinical microbiologists and many clinicians feel that the laboratory reports should be restricted to reporting *in vitro* results and should not suggest susceptibility based on published or generally accepted patterns of susceptibility. The assumption is that clinicians know, or should know, this information. Unfortunately, often they do not and treatment is based exclusively on the reported *in vitro* test results. I believe that the report should contribute to the most effective treatment of the patient. This may mean including guidelines for clinicians that are not derived exclusively from *in vitro* testing, As a clinical pathologist, I view it as my role as a consultant to produce a report that gives the clinician all the information that is needed to institute the most cost effective therapy. Clinical pharmacists increasingly are being expected to provide physicians with specific recommendations for antimicrobic use. There is no reason why microbiologists cannot do the same with appropriate prior consultation and agreement with the medical staff and often also the department of pharmacy services. Most laboratories develop profiles of susceptibility that are distributed throughout the institution in pamphlet form (Figures 13 - 14). The percent susceptibilities listed there may be included in the laboratory report as an alternative to *in vitro* testing. The information can be provided more quickly than by waiting for testing of the individual isolate and at less cost. This is especially appropriate for isolates that show a high percentage of susceptibility to the antimicrobics that are most often used clinically. We have used this approach for many years for reporting the susceptibility of anaerobes and *Staphylococcus* not aureus (Figure 15). These isolates have been collected and periodically tested to establish the profiles. One may question why not test them when isolated, instead of batching if they are all going to be tested eventually? Reporting the profile usually allows more rapid reporting, and not all isolates need to be tested to develop a profile. Reporting also may be based on published antimicrobial profiles of susceptibility (Figure 16). We have found that some physicians do not understand the percent data that is reported. Currently we are reporting "S" to an antimicrobic if the profile for the susceptibility is >55% (Figure 17). Going one step further when results of susceptibility testing of mixed cultures are reported, it might be helpful to clinicians if the activity be reported of all antimicrobics that might be used for single drug therapy for the infection. Antimicrobics such as cefotaxime, augmentin and unasyn are effective against *H. influenzae*, but their susceptibility is not usually reported because they are not first choices for the treatment of infections caused by this isolate alone. Figures 7 and 8 illustrate such a report. There is some risk that this might encourage physicians to treat a patient for every isolate that is listed. We believe that reporting of selected individual antimicrobics for each isolate does not diminish that possibility. Actually, it increases the probability that the physician will use combination therapy. If the physician is going to treat the patient for all the isolates reported, then it is preferable to encourage use of single antimicrobic therapy.

Damron and associates described complete identification and susceptibility testing on only random isolates from urine specimens from patients in a nursing home.[13] The data was used to determine the most common pathogens and their susceptibility profiles. Based on this information the most appropriate antimicrobial agents were then used for patients developing signs of urinary tract infection. This probably leads to more rapid initiation of cost-effective therapy. The cost of antimicrobial therapy and reports of misuse will continue to prompt creative approaches to reporting of susceptibility data that will optimize antimicrobial therapy and minimize cost. We established a program for recommending antimicrobial therapy based on the results of the examination of the Gram stained direct smear of sputum specimens.[15] The recommendations were issued as a part of the bacteriology report and were updated following preliminary examination of cultures and

CRITERIA FOR ESTABLISHING SUSCEPTIBILITY IN VITRO

AEROBIC BACTERIA

	Oral or minimum parenteral dose. In-vitro susc.(S) (mcg/ml)	Maximum dose In-vitro resist.(R) (mcg/ml)
Ampicillin[a]	≤ 8	≥ 16
Ampicillin[b]	≤ 8	≥ 32
Cefazolin	≤ 8	≥ 32
Cefotaxime	≤ 8	≥ 64
Cefotetan	≤ 16	≥ 64
Ceftazidime	≤ 8	≥ 32
Ciprofloxacin	≤ 1	≥ 4
Clindamycin	≤ 0.5	≥ 4
Cotrimoxazole[c]	≤ 40	≥ 80
Gent-Tobramycin	≤ 4	≥ 16
Nafcillin	≤ 2	≥ 4
Unasyn	≤ 8/4	≥ 32/16
Vancomycin	≤ 4	≥ 32

a = Enterococcus only. Isolates inhibited by ≤8mcg/ml from urine will be reported "S". Such isolates from other sites will be reported with the message that an aminoglycoside also should be used.
b = Enterobacteriaceae only.
c = Sulfamethoxazole + Trimethoprim (Bactrim or Septra).

ANAEROBIC BACTERIA

	Maximum Parenteral Dose In-vitro Susc. (S) (mcg/ml)
Cefotetan	≤ 32
Chloramphenicol	≤ 16
Clindamycin	≤ 2
Metronidazole	≤ 16
Unasyn	≤ 16/8

Peak blood levels are usually 2-4 times higher than the values shown.

AMP — Ampicillin
CFZ — Cefazolin
CHL — Chloramphenicol
CIP — Ciprofloxacin
CLI — Clindamycin
COT — Cotrimoxazole (TMP/SMX)
GEN — Gentamicin
MET — Metronidazole
NAF — Nafcillin
TAN — Cefotetan
TAX — Cefotaxime
TAZ — Ceftazidime
TOB — Tobramycin
UNA — Unasyn
VAN — Vancomycin

HARTFORD HOSPITAL

Division of Microbiology
Department of Pathology

ANTIMICROBIAL
SUSCEPTIBILITY
GUIDE

Period ending June, 1991

Figure 13. Pamphlet distributed to medical staff semiannually to guide choice of antimicrobial therapy.

ANTIBIOTIC SUSCEPTIBILITY PROFILE
Numbers are Percent Susceptible*
Period ending June, 1991

	AMP	CFZ	CIP	COT	GEN	TAN	TAX	TAZ	TOB	UNA
Acinetobacter	a		82	55	32		32	82		100
Citrobacter			92	100	88		88	76		
Enterobacter			94	87	86	53	68	68		
E. coli	81	97	100	95	98	100	100			94
Haemophilus	80			100		100	100	100		100
Klebsiella pneumoniae		88	94	94	94	98	99			
Morganella			100	100	96	100	100			
Proteus mirabilis	91	92	99	94	97	96	98			96
Proteus vulgaris				76	96	84	60	92		
Pseudomonas aeruginosa			83		58			81	85	
Serratia			96	96	100	84	100	100		

	AMP	CFZ	CLI	NAF	VAN	UNA
Enterococcus	87[b]				100	
Staph. aureus		90	87	90	100	90
Staph. not aureus		76	84	66	100	66

	CHL	CLI	MET	TAN	UNA
Bacteroides fragilis c	100	87	100	100	100
B. fragilis group[d] c	100	78	100	61	100
B. species c	100	96	100	94	100
Fusobacterium nucleat c	100	100	100	100	100

* = Based on H.H. testing and published data.
a = If percent not listed, drug not useful, or not first line chlice.
b = Therapy should include an aminoglycoside for isolates other than from urine.
c = Wexler, et al. AAC 35:1219-1224, 1991.
d = B. thetaiotaomicron, B. distasonis, B. vulgatus, B. ovatus, B. uniformis.

Figure 14. Reverse of pamphlet shown in figure 13.

```
11/29 0500 WOUND
12/03          DIRECT SMEAR - Q3
                 CELLS SEEN                       BACTERIA IN SMEAR
                 NEUTROPHILS - MANY               STAPH
                 SQUAM - NONE
               CULTURE - FINAL REPORT
                 STAPHYLOCOCCUS NOT AUREUS
                   CFZ - 76%, CLI - 84%, NAF - 66%
                 PERCENT SUSCEPTIBILITIES ARE BASED ON
                 PUBLISHED DATA AND NOT ON TESTING OF THIS ISOLATE.
                 PLEASE CONSULT MICROBIOLOGY FOR TESTING
```

Figure 15. Reporting of susceptibility of *Staphylococcus* not *aureus* based on percentage in profile. See Table 13 for key to abbreviations.

again were updated following final identification and susceptibility testing (See Figure 18). This was based on an algorithm developed by the Infectious Disease Service in our institution that resulted in a recommendation for a single parenteral and a single oral antimicrobic including the dose and frequency of administration. Physicians followed the recommendations only about 15% of the time. One reason for the poor compliance was the existence of multiple alternative acceptable therapies. Another was that treatment orders are written at the same time as orders are written for collection of cultures. Our reported results of examination of the Gram stain direct smear with initial recommendations for therapy could be retrieved on a video display terminal on the unit within 30 minutes of submitting the specimen, but this was still too late to influence the primary therapy orders. We have observed that clinicians often do not review the results of culture and antimicrobial susceptibilities following initiation of therapy unless treatment appears to be unsuccessful. Thus, in many cases the clinicians were never aware of the recommendations that had been issued in our reports. It raises serious questions about the utility of "rapid reporting". It is unlikely that anything that is reported will influence antibiotic order writing unless the information is available in guiding antimicrobial therapy elsewhere in this book. Few

```
11/29 0500 WOUND
12/03          DIRECT SMEAR - Q3
                 CELLS SEEN                       BACTERIA IN SMEAR
                 NEUTROPHILS - MANY               BACTEROIDES-HEMOPHILUS
                 SQUAM - NONE                     ENTERICS
               CULTURE - FINAL REPORT
                 E. COLI
                   AMP - S, CFZ - S, CIP - S, COT - S, GEN - S,
                   TAN - S, UNA - S
                 BACTEROIDES FRAGILIS
                   CLI - 87%, MET - 100%, TAN -100%, UNA - 100%
                 PERCENT SUSCEPTIBILITIES ARE BASED ON
                 PUBLISHED DATA AND NOT ON TESTING OF THIS ISOLATE.
                 PLEASE CONSULT MICROBIOLOGY FOR TESTING
```

Figure 16. Reporting of susceptibility of *Bacterioides fragilis* based on percentage in profile that is based on published data and not on testing of locally isolated strains. See table 13 for key to abbreviations.

```
11/29 0500 RESP. - SEC - LEUKENS
12/03          DIRECT SMEAR - Q3
                 CELLS SEEN                    BACTERIA IN SMEAR
                 NEUTROPHILS - MANY            BACTEROIDES-HEMOPHILUS
                 SQUAM - NONE                  ENTERICS
               CULTURE - FINAL REPORT
                 E. COLI
                   AMP - S, CFZ - S, CIP - S, CLI - R, COT - S, GEN - S,
                   MET - R, TAN - S, UNA - S
                 BACTEROIDES FRAGILIS
                   AMP - R, CFZ - R, CIP - R, CLI - I, COT - R, GEN - R,
                   MET - S, TAN - S, UNA - S
                 S-I-R MAY BE BASED ON LOCAL EXPERIENCE OR
                 PUBLISHED DATA.  PLEASE CONSULT MICROBIOLOGY
                 FOR TESTING
```

Figure 17. New version of report displayed in figure 16. Results for all antimicrobics to be listed for either isolate are listed for both to facilitate single antimicrobic therapy. Percent profile data used in report in figure 16. have been converted to S-I-R (see text). See table 13 for key to abbreviations.

hospitals will have the computer capability to provide recommendations for empirical antimicrobial selection upon inquiry prior to initiating therapy. Algorithms can be distributed to provide this information, and a number of therapeutic guidebooks have found their way into the pockets of house officers. It is likely that refinement of laboratory ordering systems will gradually make it attractive for physicians to interact directly with the system rather than write orders in a book to be entered into the system by a clerk. One of the advantages could be a display of recommended antimicrobics to use for lower respiratory infection at the time that submission of a sputum specimen is logged into the system. Before such interaction becomes available or acceptable to physicians, interaction with a pharmacy information system could prevent ordering of antimicrobics that do not appear in the predetermined empirical list of choices. Similarly, the pressure to reduce inappropriate costly use of antimicrobics should contribute to development of "alerts" such as have been described by Evans that will notify pharmacists, physicians, or both of discrepancies between therapeutic orders and *in vitro* test results.

Continuing attempts to reduce cost and improve effectiveness of therapy have led in our institution to experimental consideration of use of premixed antimicrobic combinations, frequently of oral agents, prepared in the hospital pharmacy. Some of these would yield substantially less costly therapy. Novel and creative naming of these agents would be a challenge. Table 12 illustrates the relative cost of some combinations of inexpensive agents that might be used compared to commonly used antimicrobic combinations that are ordered. Implementation of such a practice would offer a new challenge to the microbiologist. How would susceptibility be reported for a premixed combination of agents? This could probably be established by reporting susceptibility only when this has been observed *in vitro* for one of the two antimicrobics, based on locally collected profiles or published profile data. Alternatively, there may be *in vitro* or published evidence that synergy yields a susceptible result which would not be observed with either of the antimicrobics alone. Hopefully, such an approach will not lead to the necessity for *in vitro* synergy testing in the clinical laboratory to provide routine susceptibility tests reports! Clearly, batch testing or dependance on published data would be essential to avoid unnecessary added laboratory cost.

Table 12. Comparison of Cost of Broad Sepectrum Antimicrobics and Combinations of Older Less Expensive Agents.

<u>Single agent</u>		Cost/day ($)
Imipenem		82
Cefoxitin		38
<u>Marketed combined agent</u>		
Ampicillin/sulbactam		50
Ticarcillin/clavulanic acid		48
<u>Hospital Prepared Combination</u>	<u>Novel name</u>	
Cefotaxime/metronidazole	"Metrotax"	35
Cotrimoxazole*/clindamycin	"Zolemycin"	26
Cotrimoxazole/metronidazole	"Cotrizole"	18

Effective combinations of antimicrobics may be mixed in the hospital pharmacy prior to administration with reduction in cost and provide efficacy equal to more expensive single or marketed combination agents. Reporting of antimicrobic susceptibility would be a challenge.

---

* Trimethoprim/sulfamethoxazole
\# proposed locally contrived name for combination

```
08/06 1200 RESP. - SEC - LEUKENS
08/08          DIRECT SMEAR - Q3
                 CELLS SEEN                  BACTERIA IN SMEAR
                 NEUTROPHILS - MANY          BACTEROIDES OR HEMOPHILUS
                 SQUAM - NONE                ENTERICS
               CULTURE - FINAL REPORT
                 ENTEROBACTER CLOACAE
                 COT - I, GEN - S, TAX - S, TAZ - S
                 ACINETOBACTER CALCOACETICUS
                 COT - I, GEN - S, TAX - S, TAZ - S-
                 HEMOPHILUS INFLUENZAE
                 AMP - S, CHL - S
                 BETA LACTAMASE NEGATIVE
               RECOMMENDED THERAPY           DOSAGE
                        ALT 1     ALT 2
               08/06    TAX-C     COT-G      C= 1 G Q8H
               08/07    TAZ-C     COT-G      G= 3 AMP Q8H
               08/08    TAX-C     TAZ-C
               RECOMMENDATIONS ARE BASED ON AN ALGORITHM DEVELOPED BY THE
               DIVISION OF INFECTIOUS DISEASES AND PHARMACY.
```

Figure 18. Report including recommendations for both an oral and a parenteral antimicrobic with dosage based on gram stained direct smear. Recommendations were updated based on preliminary and final culture results including susceptibility testing. See Table 13 for key to abbreviations.

CONCLUSION

The cost of antimicrobial therapy substantially exceeds the cost of producing the bacteriologic data to guide it. Innovative approaches are needed to minimize unnecessary laboratory cost, but more importantly, to guide physicians in the most appropriate choices for therapy. Only those specimens should be accepted that are of good quality, uncontaminated by indigenous and colonizing flora, and from sites at which treatment of infection requires the guidance of *in vitro* test results. Only those isolates should be tested that are unpredictable in susceptibility, yield accurate results, and are safe to test. The antimicrobics to be tested and the results to be reported should be carefully reviewed between the laboratory, medical staff and the pharmacy service. Reporting need not be restricted to *in vitro* test results. Reports may include predictable susceptibility or resistance based on locally collected or published data. Reports should encourage the use of single antimicrobic therapy and not create a confusing array of results that may lead to a choice of multiple agents. Information systems must be developed that will lead to efficient interaction with physicians to provide guidance in antimicrobic therapy. In the interim, physicians should be given algorithms to guide initial empiric therapy. Systems should be in use to detect situations in which multiple antimicrobials may be inappropriately used because laboratory reporting of multiple isolates has suggested that a complex mixed infection may exist. Institutional regulations should permit pharmacists and/or selected infectious disease specialists to counsel physicians in such cases. Ultimately, computers may conduct most of the data collection and compilation to minimize the amount of time spent on this activity. Until then, detection of inappropriate antimicrobic use will depend on a strong interaction between the laboratory, pharmacy services, and selected representatives of the medical staff.

REFERENCES

1. L.D. Edwards, S. Levin, R. Balastra, Ordering Patterns and Utilization of Bacteriologic Culture Reports, Arch Intern Med. 132:678-682 (1973).
2. S.L. Barriere, Controversies in Antimicrobial Therapy: Formulary Decisions on Third-Generation Cephalosporins, Am J. Hosp Pharm. 43:625 (1986).
3. M.D. Parr, L.A. Hansen, W.W. Waite, Computer Program for Comparing Total Costs of Intravenous Antibiotic Regimens, Am J Hosp Pharm 43:2189 (1986).
4. R. Preissler, Personal Communication
5. R.C. Bartlett, Cost Containment in Microbiology, Clin Lab Med. 5:761-791 (1985).
6. R.C. Bartlett, C. Rutz, Processing Control and Cost in Bacteriology, Am J Clin Pathol 74:287-296 (1980).
7. A.L. Barry, P.B. Smith, M. Tuck, Cumulative Techniques and Procedures in Clinical Microbiology, Laboratory Diagnosis of Urinary Tract Infections, CUMITECH 2 (1975).
8. R.C. Bartlett, N. Treiber, Clinical Significance of Mixed Bacterial Cultures of Urine, Am J Clin Pathol 82:319-322 (1984).
9. J.M. Swenson, B.C. Hill, C. Thornsberry, Screening Pneumococci for Penicillin Resistance J Clin Microbiol 24:749-752 (1984).
10 D.M. Jones, E.M. Sutcliffe, Meningococci with Reduced Susceptibility to Penicillin, Lancet 335: 863-864 (1989).
11. J.A. Waitz, G.V. Doern, S.M. Finegold, Methods for Dilution Antimicrobial Susceptibility Tests for Bacteria that Grown Aerobically, NCCLS Doc. M7-A2, Vol. 10

12. P.D. Ellner, H.C. Neu, The Inhibitory Quotient: A Method for Interpreting Minimum Inhibitory Concentration Data, JAMA 246:1575.

13. D.J. Damron, J.W. Warren, G.R. Chippendale, Do Clinical Microbiology Laboratories Report Complete Bacteriology in Urine from Patients with Long-Term Urinary Catheters, J Clin Microbiol 24:400-404 (1986).

AREAS OF RECENT EMPHASIS OF THE NATIONAL COMMITTEE FOR CLINICAL LABORATORY STANDARDS SUBCOMMITTEE ON ANTIMICROBIAL SUSCEPTIBILITY TESTING

James H. Jorgensen

The University of Texas Health Science Center
San Antonio, Texas

INTRODUCTION

The National Committee for Clinical Laboratory Standards (NCCLS) is an organization of volunteer members representing the views of practicing clinical laboratorians, industry (both pharmaceutical and diagnostic), and government agencies. The subcommittees of the NCCLS encompass virtually all areas of clinical laboratory sciences, e.g., clinical chemistry, hematology, immunology, clinical microbiology, as well as a subcommittee which deals with broad general areas such as laboratory safety and writing of procedure manuals. However, the activities with which clinical microbiologists are most likely to be familiar are the publications emanating from the Antimicrobial Susceptibility Testing Subcommittee of the NCCLS. This subcommittee has responsibility for documents which describe the methodology for performing and interpreting antimicrobial susceptibility tests for aerobic bacteria by disk diffusion[3], by broth or agar dilution[2], and the testing of anaerobic bacteria by various methods[1]. These, like all NCCLS documents, are updated periodically (usually every three years). However, an even more frequently published document (NCCLS M100) is a compilation of all of the interpretive tables from the three susceptibility testing standards (but without the procedures text) and may be published as often as once per year.

The third edition of Performance Standards for Antimicrobial Susceptibility Testing (M1000-S3) represents the most current information provided by the Subcommittee as of this writing in the Fall of 1991.[4] A number of changes are included in this new document such as testing and reporting criteria for several new antibiotics, refinement of the language of several important explanatory foot-notes to the tables, and modification of interpretive or quality control criteria based upon newly emerging antimicrobial resistance mechanisms or testing problems brought to the attention of the subcommittee by clinical microbiology laboratories. All new

Antimicrobial Susceptibility Testing, Edited by
J.A. Poupard *et al.*, Plenum Press, New York, 1994

information published in the tables for the first time or modifications made to the tables since their last printing can be quickly recognized since the changes are indicated by boldface type.

Table 1 of M2 and M7 (Suggested Groupings of Antimicrobial Agents that Should be Considered for Routine Testing and Reporting by Clinical Microbiology Laboratories) has continued to undergo refinement in the placement of various antimicrobial agents into four groups (Groups A-D) based upon whether it is recommended that they be tested routinely or tested only under special circumstance. The criteria used by the subcommittee in decisions regarding placement of antimicrobial agents in the four groups has been detailed in a footnote to the table. The placement of agents in the different columns and rows should be viewed by clinical microbiologists as representative of the majority opinion of the subcommittee on a selective reporting scheme for those laboratories which wish to pursue that objective. Many hours were spent by the Subcommittee discussing the placement of antimicrobial agents into the four categories. The composition of the subcommittee includes not only clinical microbiologists, but also infectious diseases clinicians, pharmokineticists, and pharmaceutical industry and FDA regulatory representation. Thus, there is an opportunity for expression of widely differing views and a responsibility to develop a consensus viewpoint through compromise. The subcommittee is hopeful that the advice provided in Table 1 can be a useful starting point for local decisions regarding the specific testing and reporting needs of each institution. There are also revisions to important footnotes to Table 1 such as the suggestion that cephalothin may be tested as a representative of the first generation cephalosporins, but may not accurately predict susceptibility of *Enterobacteriaceae* to cefazolin. In addition, the footnote which urges the reporting of methicillin-resistant staphylococci as also resistant to other beta-lactams has been clarified.

Antimicrobial agents are arranged in Table 2 of M2 and M7 (zone size and MIC interpretations) according to drug classes rather than alphabetically for the first time. The goal of this change is to emphasize the antimicrobial "families" to which various agents belong such as penicillins, cephalosporins, aminoglycosides, quinolones, etc. It is hoped that this arrangement of agents in the interpretive table will familiarize microbiologists with the classes to which the various drugs belong. The subcommittee looks forward to comments from users regarding the usefulness of this change. New antibiotics included in Table 2 for the first time are cefmetazole, loracarbef (Table 2 of M2 but not M7), teicoplanin, and temafloxacin. There has been a refinement in the zone size interpretive criteria for ampicillin/sulbactam (zones reduced by 2 mm) and mezlocillin (zones also reduced by 2 mm when testing *Pseudomonas spp.*), and a major change in interpretive zone sizes for vancomycin when testing enterococci. The changes are indicated in Table 1 of this article. The change in vancomycin disk interpretive criteria came about in response to the recent development of varying degrees of vancomycin-resistance among enterococci, including lower level resistance not detected reliably by the previous NCCLS zone size breakpoints.[5] The revised zone sizes included in this document for interpretation of vancomycin disk tests of enterococci will provide more accurate detection of resistance among contemporary enterococcal strains. However, it is recommended that a vancomycin MIC should be performed on any enterococcal isolate from a serious infection if the disk test result is intermediate. The vancomycin interpretive zones have not been changed for other gram positive bacteria and the vancomycin MIC breakpoints have not been revised.

Table 1. Zone size interpretive breakpoints which have been revised in NCCLS M100-S3.

Antibiotic	Revised Zones	Previous Zones
Ampicillin/sulbactam	<11, 10-14, >15	<13, 14-16, >17
Mezlocillin (when testing *Pseudomonas*)	<15, - , >16	<17, -, >18
Vancomycin (when testing *enterococci*)	<14, 15-16, >17	<9, 10-11, >12

Table 2. Antibiotics for which routine quality control of Haemophilus test medium with *H. influenzae* ATCC 49766 is now recommended.

ANTIBIOTIC	ACCEPTABLE CONTROL RANGE	
	MICs (μg/ml)	Zones (mm)
Cefaclor	1 - 4	25 - 31
Cefamandole	0.25 - 1	29 - 37
Cefonicid	0.06 - 0.25	30 - 38
Cefuroxime	0.25 - 1	28 - 36
Loracarbef	0.5 - 2	26 - 32

Table 3. Revised MIC (μg/ml) quality control ranges for testing aminoglycosides with *Pseudomonas aeruginosa* ATCC 27853 using calcium adjusted Mueller-Hinton broth.

ANTIBIOTIC	Revised Range	Previous Range
Amikacin	0.5-8	2-8
Gentamicin	0.25-4	1-4
Netilmicin	0.5-8	2-8
Tobramycin	0.12-2	0.5-2

The subcommittee has continued to expand the area of testing of fastidious bacteria. Table 2A (breakpoints for *Haemophilus spp.*) of M2 and M7 includes the addition of lomefloxacin interpretive criteria, while Table 2A of M2 indicates a revised breakpoint (smaller zone sizes) for determination of cefixime susceptibility of *Haemophilus* species by the disk test. Table 2B of M2 and M7 includes new MIC and disk interpretive criteria for a number of additional beta-lactam and quinolone antibiotics when tested against *N. gonorrhoeae*. Table 3A incorporates a new strain of *H. influenzae* (ATCC 49766) for use in quality control testing of five cephalosporins. This alternative strain was identified in response to the experience that *H. influenzae* ATCC 49247 provided conflicting results with certain cephalosporins. The new strain is now recommended for routine quality control of cefaclor, cefamandole, cefonicid, cefuroxime, and loracarbef. The acceptable ranges of zones for the new strain are indicated in Table 2 of this article. A new Table 3B in M2 and M7 includes quality control values for an expanded array of agents for testing *N. gonorrhoeae*. Table 7 of M7 indicates that Haemophilus test medium may be used for broth microdilution MIC testing of *Streptococcus pneumoniae* as an alternative to the use of lysed horse blood-supplemented Mueller-Hinton broth; although some differences in MICs may occur with certain drugs.

Another important quality control change has resulted from the lowering of the recommended cation (magnesium and calcium) content of Mueller-Hinton broth by approximately one-half. The cation content has been "adjusted" in order to provide comparable aminoglycoside MICs with *Pseudomonas aeruginosa* irrespective of whether the testing was performed by broth or agar dilution methodology. The lowered cation concentrations have necessitated a revision to the acceptable MIC ranges for four aminoglycosides when testing the control strain *P. aeruginosa* ATCC 27853. As can be seen in Table 3, the MIC ranges for each drug have been expanded by two log2 dilutions only on the lower end of the acceptable ranges. As more data become available, it may be possible to make further refinements to these ranges.

The use of MICs indicative of susceptibility with anaerobic bacteria which are higher than MICs indicating susceptibility to aerobes have been reaffirmed in the tables which have been excerpted from Table 1 of NCCLS document M11[1]. In general, the MIC which corresponds to the moderately susceptible category in M7 (aerobes 2) represents the susceptible category for anaerobic bacteria. This is based in part on the assumption that serious infectious are treated with maximum dosages of antibiotics, and therefore blood levels should reach those maximally attainable with the various agents.

In summary, the addition and changes which appear in the latest NCCLS susceptibility document should be viewed as evolutionary rather that revolutionary. Several new antibiotics have been added to the various tables of the document, and interpretive criteria for a few drugs have been revised due to recognition of new bacterial resistance mechanisms, or corrected due to more recent or more extensive data. The virtue of the NCCLS consensus development process is the ability to add new information and update older information as the need arises. Inherent in this process is the participation of clinical microbiologists in offering comments regarding the content of the most recently published NCCLS documents.

The use of MICs indicative of susceptibility with anaerobic bacteria which are higher than MICs indicating susceptibility to aerobes have been reaffirmed in the tables which have been excerpted from Table 1 of NCCLS document M11[1]. In general, the MIC which corresponds to the moderately susceptible category in M7 (aerobes 2) represents the susceptible category for anaerobic bacteria. This is based in part on the assumption that serious infectious are treated with maximum dosages of antibiotics, and therefore blood levels should reach those maximally attainable with the various agents.

In summary, the addition and changes which appear in the latest NCCLS susceptibility document should be viewed as evolutionary rather that revolutionary. Several new antibiotics have been added to the various tables of the document, and interpretive criteria for a few drugs have been revised due to recognition of new bacterial resistance mechanisms, or corrected due to more recent or more extensive data. The virtue of the NCCLS consensus development process is the ability to add new information and update older information as the need arises. Inherent in this process is the participation of clinical microbiologists in offering comments regarding the content of the most recently published NCCLS documents.

REFERENCES

1. National Committee for Clinical Laboratory Standards 1990, Methods for antimicrobial susceptibility testing of anaerobic bacteria. Approved standard M11-A2 National Committee for Clinical Laboratory Standards, Villanova, PA.
2. National Committee for Clinical Laboratory Standards 1990, Methods for dilution antimicrobial susceptibility tests for bacteria that grow aerobically. Approved Standard M7-A2. National Committee for Clinical Laboratory Standards, Villanova, PA.
3. National Committee for Clinical Laboratory Standards, 1990, Performance standards for antimicrobial disk susceptibility tests. Approved standard M2-A4. National Committee for Clinical Laboratory Standards, Villanova, PA.
4. National Committee for Clinical Laboratory Standards, 1991, Performance standards for antimicrobial susceptibility testing. Third informational supplement M100-S3. National Committee for Clinical Laboratory Standards, Villanova, PA.
5. Swenson, J.M., B.C. Hill, and C. Thornsberry, Problems with the disk diffusion test for detection of vancomycin resistance in enterococci, J. Clin. Microbiol. 27:2140-2142 (1989).

NON-TRADITIONAL APPROACHES FOR QUALITY CONTROL OF ANTIMICROBIAL SUSCEPTIBILITY TESTS

Janet Hindler

UCLA Medical Center, Los Angeles, California

INTRODUCTION

Quality control is an essential part of any clinical laboratory procedure and can simply be defined as the series of steps that are taken to make certain test results are accurate and reproducible. Antimicrobial susceptibility tests, like many other microbiology tests are somewhat unique in that they depend on growth (or no growth) of bacteria which is often unpredictable and sensitive to changes in the environment. Therefore, it is essential that technical variables in the test system are well standardized and controlled and that personnel performing the tests thoroughly understand the components of the test system. This must all be considered when developing a program for quality control of antimicrobial susceptibility tests.

Traditional Approaches

As part of an antimicrobial susceptibility testing QC program, laboratories must regularly test reference strains with known antimicrobial susceptibility profiles. A familiar and practical approach for inclusion of such QC reference strains that is used by most clinical laboratories is described in detail by the National Committee for Clinical Laboratory Standards (NCCLS).[16,17,18] The strains currently recommended for QC of routine disk diffusion and dilution tests are listed in Table 1.

When results with these reference strains agree with predefined criteria, the test system is considered to be in control. Parameters of the test system that are controlled to various degrees by testing the QC strains are:

- antimicrobial potency
- test medium
- pH
- cations (Ca^{++} Mg^{++})
- thymidine
- medium composition
- growth factors
- agar depth (disk diffusion)
- incubation conditions (temperature, etc.)
- standarization of inoculum and inoculation technique
- batch contamination of media/supplies
- instrument(s) functioning
- measurement of endpoints

Antimicrobial Susceptibility Testing, Edited by
J.A. Poupard *et al.*, Plenum Press, New York, 1994

Table 1. Strains Currently Recommended for QC of Routine Disk Diffusion and Dilution Tests

QC Strain	Primarily used for QC testing of:
E. coli ATCC 25922 [a]	Gram (-) drugs
S. aureus ATCC 25923 (*B*-lactamase -) [a]	Gram (+) drugs - disk diffusion; GC - disk diffusion
S. aureus ATCC 29213 (*B*-lactamase +) [a]	Gram (+) drugs - dilution tests; GC - dilution tests
P. aeruginosa ATCC 27853	aminoglycosides (+ others); pH of system
E. coli ATCC 35218	*B*-lactam/*B*-lactamase inhibitor combinations
E. faecalis ATCC 29212	media, to check that it is acceptable for testing sulfonamides, trimethoprim, and trimethoprim/sulfamethoxazole; also for other drugs (dilution tests)
H. influenzae ATCC 49247	*H. influenzae*
H. influenzae ATCC 49766	*H. influenzae* (select cephalosporins)
N. gonorrhoeae ATCC 49226	*N. gonorrhoeae*

[a] may include for supplemental testing of other drugs that are active against these strains.

As would be expected, there are limitations of testing the standard QC reference strains in controlling results when testing patient isolates. Some of the parameters not readily controlled are:

- individual antimicrobial/organism test problem (disk, well, tube)
- sporadic contamination of broth systems
- sporadic instrument malfunction
- NaCl content for oxacillin broth tests with staphylococci
- subjective reading of "difficult" endpoints, (e.g. "trailing", "skipped wells")
- interpretation of results, use of appropriate interpretive criteria
- transcription errors
- individual technical errors

Additionally, often patient isolates are tested which are considerably different from the recommended QC reference strains whereby testing of these is only modestly controlled by performance testing of the suggested QC reference strains.

Some of these include:

- non-enterococcal streptococci
- *Corynebacterium spp.*
- other fastidious organisms
- organisms with "unusual" growth characteristics
    - mucoid *Pseudomonas aeruginosa*
    - dwarf *E. coli*

- organisms with unusual resistance characteristics
    - oxacillin-resistant staphylococci
    - organisms with inducible resistance
- organisms for which resistance is uncommon (e.g., ciprofloxacin-resistant *Enterobacteriaceae*)

In addition to regular testing of QC reference strains, participation in proficiency testing programs can be considered another "traditional" QC measure for antimicrobial susceptibility tests. Assuming that proficiency survey samples are handled in a manner similar to patient samples, satisfactory performance when testing these will provide further assurance that your laboratory is generating accurate results.

A recent College of American Pathologists (CAP) proficiency survey report[12] noted the following (Table 2) regarding disk diffusion QC testing over the past decade from participants in their Bacteriology Proficiency Survey program:

Table 2. Disk Diffusion QC: Summary of CAP Proficiency Survey Report

Parameter	1981	1984	1987	1989
laboratories performing weekly (vs daily QC) testing of reference strains	18[a]	37	67	69
laboratories with ≤2% of QC out-of-control results	85	96	97	98
laboratories repeating testing of QC strains as the only corrective action for out-of-control results	52	80	74	77
laboratories following NCCLS QC guidelines	59	83	97	99

[a] percent of subscribers

Most laboratories are performing weekly QC testing of the reference strains now that this is recognized as an acceptable practice by NCCLS and all of the major accrediting agencies. More laboratories than ever before are obtaining satisfactory results with the QC strains for >98% of QC tests and errors usually correct upon repeat testing. Nearly all of the laboratories that participate in the CAP Bacteriology Proficiency Survey program follow the QC testing guidelines specified by NCCLS. These data clearly show that QC testing of the reference strains does not represent a particular problem for clinical microbiology laboratories.

It is obvious that satisfactory performance testing of the recommended QC reference strains is extremely valuable, but this alone does not ensure consistently accurate and reproducible results upon testing patient isolates. Thus, additional measures must be included in an antimicrobial susceptibility testing QC program.

SUPPLEMENTAL STRATEGIES FOR QUALITY CONTROL OF ANTIMICROBIAL SUSCEPTIBILITY TESTS

Inclusion of Supplemental QC Strains

In addition to utilizing the NCCLS recommended QC reference strains, a laboratory may elect to periodically augment QC of the test system by testing in-house selected organisms. Since detection of oxacillin-resistant *S. aureus* (ORSA) is subject to minor alterations in test conditions, including an ORSA isolate would seem reasonable. Similarly, there is significant concern about "rapid systems" detecting beta-lactam resistance in *Enterobacteriaceae* and a laboratory that is using a rapid system might want to include, for example, an isolate of *Enterobacter cloacae* that is resistant to ampicillin and several cephalosporins.

One source for supplemental QC strains would be proficiency surveys, where coordinators of these programs often select organisms that provide significant challenges for clinical laboratories.[12] The attractive features about using proficiency survey organisms are that 1) all accredited clinical microbiology laboratories obtain these and 2) the information provided with the scoring is often substantial and data reflect the performance of many laboratories using a variety of test systems. Alternatively, supplemental QC organisms could be procured from other microbiologists or the American Type Culture Collection (ATCC).[1] With these, laboratories are often faced with starting from scratch to establish any QC limits.

Some manufacturers of commercial systems are recommending user laboratories test additional strains that they have specifically selected for QC of their test system. Company representatives should provide these organisms as well as any other organisms with specific antimicrobial susceptibility characteristics that you might wish to examine. It is important to note that the resistance characteristics of some strains that might be selected as "in-house" QC strains might not always be stable and this must be considered when a discrepancy is noted. A practical strategy for periodic testing of supplemental QC strains would be to use them to:

- make certain the test system can reliably detect a specific type of resistance (e.g., ORSA)
- train students and new technologists
- periodically check proficiency of technologists performing testing
- trouble-shoot specific problems
- establish confidence in a new test system used in your laboratory

Proficiency of Technologists

Because significant judgment is required when performing antimicrobial susceptibility tests, a primary focus should be on making certain technologists are proficient in all aspects of the testing which includes becoming knowledgeable in how to verify the accuracy of individual patient results. Checklists customized to your laboratory's needs as shown in Appendix 1 can be periodically distributed to technologists for self-assessment to make certain the most critical aspects of testing are understood. Technologists should be encouraged to seek assistance when they recognize that their knowledge in specific areas is lacking.

Proficiency in the manual reading of MIC and disk diffusion tests can be assessed by having all technologists read the MICs or measure the zones of inhibition for the same MIC tray or Kirby Bauer plate. MIC results should agree within +/- one two-fold dilution[2] and inhibition zones should be within 2 mm.[3] Periodic inservice sessions for laboratory staff are essential. Some of the most valuable resources for these sessions include the NCCLS documents and CAP proficiency survey critiques. A 45 min - 1 h session dedicated to discussing NCCLS

Table 1 (including footnotes) from NCCLS M7-A2 and M2-A4 "Suggested Groupings of Antimicrobial Agents That Should be Considered for Routine Testing and Reporting by Clinical Microbiology Laboratories" [16,17,18] will uncover a wealth of practical information. It is particularly important to review this and other portions of the NCCLS documents when new versions are released. Circulation of the CAP proficiency survey critiques and documentation that all technologists have reviewed this information will help inform technologists about current antimicrobial susceptibility testing issues and will also satisfy the requirement (proficiency survey information be reviewed with the staff) mandated by some accrediting agencies.

A mechanism for a supervisor or supervisor designee to review all antimicrobial susceptibility test results prior to release will undoubtedly identify some errors that may not be apparent to every technologist. It seems that because of the increasing amount of information needed to understand antimicrobial susceptibility results, many laboratories are identifying a dedicated individual(s) (which may be the supervisor or a technologist) to serve as the resource person for antimicrobial-related questions. This individual(s) also performs the final review of susceptibility results. Obviously, any question that cannot be comfortably handled by a supervisor or technologist must be referred to the laboratory director.

Antibiograms

An antibiogram can be defined as the overall antimicrobial susceptibility profile of a bacterial isolate to a battery of antimicrobial agents. The antibiogram is an extremely useful tool to help verify the accuracy of antimicrobial results. Additionally, the utility of the antibiogram in verifying an organism's identification must be underscored.[21] Specific bacteria often have predictable or "typical" susceptibility or resistance to particular drugs and the identity of a specific isolate can be checked against its expected antibiogram. To optimize use of antibiograms, one must become familiar with typical antibiograms.

It is important to be able to differentiate typical from atypical antibiograms. Because of numerous factors, not the least of which is the increasing use of antimicrobial agents, organisms with atypical antibiograms that are verified to be true are becoming more common. However, atypical antibiograms, where results cannot be duplicated may occur as a result of technical or clerical errors. These can often be identified providing one becomes knowledgeable in the concepts described above. Laboratories must provide mechanisms to aid technologists in the consistent recognition of atypical antibiograms and provide a strategy for handling these that will minimize the reporting of erroneous results when testing patient isolates. Computer reporting systems often have the capability to "build in" basic antibiogram check mechanisms. However, these do not preclude the need for technologists to check results that are generated.

In developing an antibiogram check program to aid technologists in consistently identifying atypical antibiograms, the following should be provided:

- descriptions of the relatedness of drugs tested (e.g. activity hierarchy)
  For example, the activity hierarchy of the three generations of cephalosporins against the *Enterobacteriaceae* is:
  3rd > 2nd > 1st

Similarly, the activity hierarchy of the aminoglycosides against the *Enterobacteriaceae* is:
amikacin > tobramycin > gentamicin = netilmicin

- descriptions of typical antibiograms for given species (e.g. the typical *S. aureus* is resistant to penicillin and susceptible to clindamycin, erythromycin, oxacillin, and vancomycin)

- informal updates to advise the staff of the prevalence of a particular "atypical" antibiogram at a given time (e.g. increased incidence of nosocomial infections due to gentamicin-resistant *Providencia rettgeri*)

Appendix 2A - 2D lists antibiograms most likely to be encountered for the respective species. These, however, represent very general guidelines and exceptions occur frequently. Additional information on antibiogram for glucose-nonfermenting gram-negative rods can be found in reference 8.

One exception to the typical hierarchy of activity of cephalosporins against the *Enterobacteriaceae* occurs with extended spectrum beta-lactamases (ESBL). These plasmid-mediated enzymes are believed to evolve from point mutations of the TEM and SHV type enzymes commonly associated with ampicillin resistance in *E. coli* and *Klebsiella pneumoniae*.[11,19] The common TEM (TEM-1) and SHV (SHV-1) enzymes do not hydrolyze third generation cephalosporins (eg cefotaxime, ceftazidime) or aztreonam. However, ESBL may hydrolyze these and other extended-spectrum beta-lactam agents to varying degrees. Imipenem and cefoxitin are generally not effected and ceftazidime is the most vulnerable of the third generation cephalosporins.[11,19] An *E. coli* isolate (D-5 1991) recently included in a CAP Bacteriology proficiency survey produced an ESBL and had the following antibiogram[5]:

ampicillin	R
cefazolin	R
cefoxitin	S
cefotaxime	S
ceftazidime	**R**
ceftizoxime	S
ceftriaxone	S

With isolates as this, laboratories that follow a selective reporting protocol (e.g., generally report 3rd generation cephalosporins only if an isolate is resistant to the 1st and 2nd generation agents tested) must report results of ALL cephalosporins tested to prevent physicians from extrapolating results inappropriately. In this example, based on the cefoxitin results the tendency would be to assume that the isolate is susceptible to third generation cephalosporins. At this time, it is not known if all cephalosporins would be clinically ineffective against isolates that produce ESBL and the incidence of isolates with ESBL is low in most parts of the United States. We must all keep abreast of developments in this area so we can modify our laboratory testing and reporting protocols, if necessary.

Occasionally, an exception to the aminoglycoside hierarchy of activity is noted in a particular isolate. Most aminoglycoside resistance is due to the production of aminoglycoside inactivating enzymes.[10] Examples of a few of the aminoglycoside inactivating enzymes produced by *Enterobacteriaceae* and the aminoglycosides that they inactivate are:

Enzyme	Gentamicin	Tobramycin	Amikacin	Netilmicin
ANT (2")	X	X		
ANT (4')		X	X	
AAC (6')-a		X	X	X
AAC (6')-b	X	X		X

We recently observed an *Enterobacter aerogenes* that was resistant to gentamicin, tobramycin, and amikacin, but susceptible to netilmicin. This isolate is currently being analyzed further, however, it is believed to possess an ANT(2") and an ANT(4'). It is not uncommon to encounter a *Serratia marcescens* that possesses an AAC (6')-a that confers resistance to tobramycin, amikacin, and netilmicin, but not to gentamicin.

The subject of antimicrobial resistance is very complex and it is impossible for clinical microbiologists to be aware of all the exceptions to "typical antibiograms" that might be encountered. This is complicated by the fact that new mechanisms of antimicrobial resistance are continually being identified. Any unusual antibiogram should be discussed with the laboratory director, who will determine whether this should be investigated further with the patient's clinician and specific researchers. Two recent summaries of current issues related to antimicrobial resistance include an article by Jacoby and Archer[10] and another by Murray.[15]

Descriptions of the types of antimicrobial resistance (or susceptibility) that when reported (or missed) are likely to impact on patient care should be repeatedly emphasized. Examples would include *P. aeruginosa* resistant to amikacin, gentamicin and tobramycin particularly from systemic infections, ORSA from any source, and penicillin-resistant or penicillin-relatively-resistant *Streptococcus pneumoniae* from CSF. Once an atypical antibiogram is identified, the next step is to determine, what, if anything should be done to verify these results. A common misconception is that verification of atypical antibiograms always necessitates repeat testing. An example of a more cost-effective approach to verify atypical antibiograms would involve the following steps:

- check for transcriptional errors
- reexamine KB plate, MIC tray, purity plate, etc. to make certain test appears satisfactory. Subtle problems may not always be detected upon initial examination of the test.
- check previous reports on the patient to determine whether this atypical antibiogram represents a repeat occurrence
- repeat susceptibility tests, identification tests, or both

In some situations it may be helpful to use an alternative test method to verify unusual results. Additionally, it may be appropriate to solicit the assistance of a reference laboratory that specializes in antimicrobial testing to verify virtually "unheard of" atypical results (e.g., vancomycin-resistant *S. aureus*). Some situations may warrant assistance from your local Public Health Department (e.g., the isolation of a *S. pneumoniae* that is resistant to penicillin with a penicillin MIC of 8 mcg/ml).

Technologists, often with supervisorial assistance, must use their judgment in determining when to report a result that requires verification by repeat testing. Obviously, we do not want to delay the report, however the consequences of reporting an erroneous result may have a greater adverse effect on patient care than a delayed report. Each situation must be evaluated on an individual basis.

In our laboratory, we have developed a list of conditions that always require verification Appendix 3. Our rationale for including some of these is based on the impact that erroneous results might have on patient care. Additionally, some of these represent situations where we have had significant problems in the past or represent situations that are very unlikely to occur.

As for clinical impact, we can look again at ORSA. All of us know that the patient infected with an oxacillin-susceptible *S. aureus* is usually treated with a penicillinase-resistant penicillin as oxacillin. In contrast, the more costly and toxic vancomycin is the drug of choice for ORSA infections.[13] Additionally, the hospitalized patient with ORSA requires isolation. The added cost, and clinical and psychosocial consequences for the patient of our reporting an ORSA are significant. Therefore, we must take maximum precautions when reporting this type of result.

We know that effective treatment of enterococcal endocarditis requires a combination of a cell wall active agent (generally ampicillin, penicillin, or vancomycin) and an aminoglycoside.[14] Increasing numbers of isolates with high level aminoglycoside resistance have been identified and these isolates are refractory to treatment with the particular aminoglycoside(s) in combination with the cell wall active agent. Reliable *in vitro* test methods are essential to detect this clinically important high level aminoglycoside resistance.

Erroneously reporting vancomycin resistance on a *S. aureus* isolated from a superficial wound may not move the general practitioner who has prescribed dicloxacillin to an outpatient that has subsequently responded favorably to this agent. However, this same information compiled into the laboratory's cumulative antimicrobial statistics may create minor chaos among the infectious disease specialists.

Observed vancomycin resistance in a "presumptive" *viridans streptococcus* should suggest further testing be performed to determine whether this isolate might be a *Pediococcus spp.*, *Leuconostoc spp.*, or *Lactobacillus spp.* These organisms morphologically resemble viridans streptococci but unlike *viridans streptococci*, *Pediococcus spp.*, *Leuconostoc spp.*, and *Lactobacillus spp.* are typically vancomycin resistant.[20]

Many of the examples stated thus far suggest specific types of resistance that should be verified. The more difficult check would be to verify "susceptible" results. Most of us would agree that it would be necessary to verify results when an ampicillin- and cephalothin-susceptible *Enterobacter cloacae* is encountered. However, in many situations there is no "clue" that suggests a susceptible result may be erroneous. Reporting erroneously susceptible results can obviously have a significant clinical impact. For example, penicillin is the drug of choice for the patient with endocarditis due to a penicillin-susceptible *viridans streptococcus*. However, the addition of an aminoglycoside would be warranted when treating this same patient, were he to have a moderately susceptible isolate.[4] To reiterate, the antibiogram check is but one component of the antimicrobial susceptibility testing QC program and probably allows us to spot potentially false resistant results better than those that are falsely susceptible. One check to minimize reporting of false susceptible results would involve making sure that growth of the organism in the test system is sufficient. It is sometimes tempting to read results even when growth on the Kirby Bauer plate or MIC tray is "light". Reliable results can only be obtained with a disk diffusion test when there is a lawn of confluent growth surrounding the zones of inhibition. Similarly, 3+ or 4+ turbidity or a large button of growth in the growth control well is needed in microdilution MIC tests.

It is important to review all results generated and not just those that are reported. For example, a *P. aeruginosa* isolate that is ampicillin susceptible but presents with an otherwise unremarkable, but very susceptible antibiogram (e.g., susceptible to aminoglycosides, mezlocillin, piperacillin, and ceftazidime) has probably been underinoculated and all results might be falsely susceptible. The dangers in reading results from tests that do not demonstrate adequate growth are obvious.

Needless to say, all of the results generated by the laboratory are important. However, it is obvious that the consequences of certain reports will have a much greater impact than others. It is therefore critically important for us to continue to obtain insight into the results that are likely to have the greatest clinical impact and act on these appropriately.

Cumulative Susceptibility Statistics

Another measure that can be taken to further ensure reporting of accurate and reproducible antimicrobial susceptibility results on patient isolates involves periodic review of cumulative susceptibility statistics which may uncover subtle inconsistencies. For example, as stated by Washington, if greater than 5% of *Enterobacter*, *Serratia*, or *Citrobacter* are susceptible to ampicillin, there is likely to be a problem with use of an insufficient inoculum.[22] Boyce and colleagues observed an increase in the number of ORSA in their patient population and discovered defective oxacillin disks. This problem went virtually undetected with routine testing of the QC reference strains.[6]

As summarized in Appendix 4 there are multiple components to an effective program for quality control of antimicrobial susceptibility tests that will ensure "quality" results when testing patient isolates. In addition to testing the QC reference strains as recommended by NCCLS, we must maintain a continual awareness of the critical parameters involved in test performance, the typical results that we would expect to encounter for a given organism, and the impact that specific results might have on patient care. Undoubtedly, it is critical that the entire microbiology laboratory staff participate in this ongoing program.

This lecture was taken in part from Reference 9.

APPENDIX 1

NAME ______________________________ DATE ______________

Technologist Proficiency Checklist for the Disk Diffusion Test

Primary indications for use; limitations of procedure
Storage, handling, characteristics of test materials and equipment

Preparation of inoculum:
"picking" bacteria
log phase-incubate 2-8 h
direct inoculum standardization-overnight colonies

Performance of test:
inoculation of MHA-within 15' of standardization, 3 directions
placement of disks-within 15' of inoculation of plate, do not relocate disks

Incubation:
16-18 h (18-24 h for slower growing organisms), 35°C non CO_2 enhanced environment, stack ≤ 5 high

Reading and interpretation of results:
reflected light (except oxacillin[1] and staphylococci)
confluent lawn
zone edges; trimethoprim and sulfonamides ($\geq$ 80% inhibition)
swarming Proteus
colonies within zone
NCCLS interpretive charts:
- definition of S, MS, I, R
- drugs with multiple interpretive criteria for different organism groups
- specific criteria for urinary tract isolates
- other NCCLS information including footnotes to tables

***Haemophilus influenzae*:**
Haemophilus Test Medium (HTM)
direct inoculum standardization
beta-lactamase test
chloramphenicol acetyltransferase test

***Streptococcus pneumoniae*:**
direct inoculum standardization
MHA-5% sheep blood
oxacillin 1 μg disk to test for penicillin susceptibility

***Neisseria gonorrhoeae*:**
- direct inoculum standardization
- GC agar base with supplement
- beta-lactamase test

***Staphylococcus spp.*:**
- MHA
- direct inoculum standardization
- oxacillin[1]-incubate 24 h 33-35°C; examine zones with transmitted light
- "clues" suggesting possible oxacillin[1]-resistant *S. aureus*
- reporting of other drugs, i.e. cephalosporins, for oxacillin[1]-resistant isolates

Trimethoprim, sulfonamides-do not test on media with blood products (except horse blood)

Reporting results:
- appropriate drugs
- indications for "presumptive" results
- indications for supervisorial assistance

QC procedures:
- routine QC
- expected accuracy and reproducibility
- antibiograms
- clinically relevant results

Reference materials:
- NCCLS documents
- Manual of Clinical Microbiology V
- Other

[1] includes all penicillinase-resistant penicillins (oxacillin, methicillin, nafcillin, cloxacillin, dicloxacillin, flucloxacillin)

APPENDIX 2A

THESE ANTIBIOGRAMS SHOULD SERVE AS GUIDELINES ONLY - EXCEPTIONS WILL OCCUR.

Enterobacteriaceae Antibiograms

ORGANISM	AMP	AMP/ SULB	CARB GRP	MEZLO GRP	1ST CEFS	CFMDL GRP	CEFOX	CEFOT	CFTAX GRP	CFPRZ	IMIP	GENT TOB NET	AMIK	CHLOR	TETRA	TMP/ SMZ	CIPRO GRP	NALID GRP	NRFLX	NITRO FURANT	POLYB
MIC Breakpoint for Susceptibility (μg/ml)	≤ 8	≤ 8	≤16	≤ 16	≤ 8	≤ 8	≤ 8	≤ 8	≤ 8	≤ 16	≤ 4	≤ 4 [a]	≤ 16	≤ 8	≤ 4	≤ 2/38	≤ 1 [b]	≤ 16	≤ 4	≤64	≤ 4
Aeromonas hydrophlia [c]	R*	R	R*	S-R	R	S-R	S-R	S-R	S	S	S	S	S	S	S	S	S	S	S	S	S
Citrobacter diversus	R*	S-R	R	S	S	S	S	S	S	S	S	S	S	S	S	S	S	S	S	S	S
Citrobacter freundii	R*	S-R	S	S	R*	S-R	R	S-R	S	S	S	S	S	S	S	S	S	S	S	S	S
Enterobacter aerogenes *Enterobacter cloacae*	R*	S-R	S	S	R*	S-R	R	S-R	S	S	S	S	S	S	S	S	S	S	S	S-R	S
Enterobacter agglomerans	S-R	S-R	S	S	S-R	S-R	S-R	S-R	S	S	S	S	S	S	S	S	S	S	S	S-R	S
E. coli 1	S	S	S	S	S	S	S	S	S	S	S	S	S	S	S	S	S	S	S	S	S
E. coli 2	R	S-R	S-R	S-R	S-R	S	S	S	S	S	S	S	S	S-R	S-R	S	S	S	S	S	S
Klebsiella spp.	R*	S-R	R*	S-R	S	S	S	S	S	S	S	S	S	S	S	S	S	S-R	S	S-R	S
Morganella morganii	R*	S-R	S	S	R*	S-R	S-R	S	S	S	S	S	S	S	S-R	S	S	S	S	R*	R*
Proteus mirabilis	S-R	S-R	S	S	S	S	S	S	S	S	S	S	S	S	R*	S	S	S	S	R*	R*
Proteus vulgaris	R*	S-R	S	S	R*	R	S-R	S	S	S	S	S	S	S	R*	S	S	S	S	R*	R*
Providencia spp.	R*	S-R	S	S	R*	S-R	S-R	S	S	S	S	S	S	S	R*	S	S	R	S	R*	R*
Salmonella spp.	S	S	S	S	S	S	S	S	S	S	S	S	S	S	S	S	S	S	S	S	S
Serratia spp.	R*	R	S	S	R*	R	S-R	S-R	S	S	S	S	S	S	R	S	S	S-R	S	R*	R*
Shigella spp.	S	S	S	S	S	S	S	S	S	S	S	S	S	S	S	S	S	S	S	S	S

CARB GRP = carbenicillin, ticarcillin
MEZLO GRP = mezlocillin, piperacillin
1ST CEFS = cefazolin, cephalothin
CFMDL GRP = cefamandole, cefonicid, cefuroxime
CEFOX = cefoxitin
CEFOT = cefotetan
CFPRZ = cefoperazone
CFTAX GRP = cefotaxime, aztreonam, ceftazidime, ceftizoxime, ceftriaxone
CIPRO GRP = ciprofloxacin, ofloxacin
NALID GRP = nalidixic acid, cinoxacin
NRFLX = norfloxacin

[a] ≤ 8 for netilmicin

[b] ≤ 2 for ofloxacin

[c] member of Vibrionaceae

* very unusual to encounter an isolate susceptible to this drug

APPENDIX 2B

THESE ANTIBIOGRAMS SHOULD SERVE AS GUIDELINES ONLY - EXCEPTIONS WILL OCCUR.

Pseudomonas – Acinetobacter - Xanthomonas Antibiograms

ORGANISM	AMP	AMP/ SULB	CARB GRP	PIPER GRP	1ST CEFS	2ND CEFS	CFTAX GRP	CFTAZ	CFPRZ	AZTRE	IMIP	GENT TOB NET	AMIK	CHLOR	TETRA	TMP/ SMZ	CIPRO GRP	NALID GRP	NRFLX	NITRO FURANT	POLYB
MIC Breakpoint for Susceptibility (µg/ml)	≤ 8	≤ 8	≤128[a]	≤ 64[b]	≤ 8	≤ 8	≤ 8	≤ 8	≤ 16	≤ 8	≤ 4	≤ 4[c]	≤ 16	≤ 8	≤ 4	≤ 2/38	≤ 1 [d]	≤ 16	≤ 4	≤ 64	≤ 4
Acinetobacter anitratus	R	S-R	S	S	R	R	S-R	S	S-R	R	S	S	S	R	S-R	S	S	S-R	S-R	R	S
Acinetobacter lwoffi	R	S-R	S	S	R	R	S-R	S	S-R	R	S	S	S	S-R	S-R	S	S	S-R	S-R	R	S
Pseudomonas aeruginosa	R*	R*	S-R	S-R	R*	R*	R	S	S	S	S	S	S	R	R	R	S	R	S	R	S
Pseudomonas cepacia	R*	R	R	S-R	R*	R*	S-R	S	S	S-R	R	R	R	S	R	S	R	R	R	R	R
Pseudomonas fluor-putida	R*	R*	R	S-R	R*	R*	R	S	S	S-R	S	S	S	R	S-R	R	S-R	R	S-R	R	S
Xanthomonas maltophilia	R	R	S-R	S-R	R*	R*	R	R	R	R	R	R	R	S	R	S	R	R	R	R	S
Pseudomonas stutzeri	S-R	S-R	S-R	S-R	R	R	S-R	S	S	S-R	S	S	S	S-R	S	S	S-R	R	S-R	R	S

CARB GRP = carbenicillin, mezlocillin, ticarcillin
PIPER GRP = piperacillin, azlocillin
CFTAX GRP = cefotaxime, ceftriaxone
CFTAZ GRP = ceftazidime
CFPRZ = cefoperazone
AZTRE = aztreonam
CIPRO GRP = ciprofloxacin, ofloxacin
NALID GRP = nalidixic acid, cinoxacin
NRFLX = norfloxacin

[a] ≤128 for Pseudomonas (CARB); ≤ 64 for Pseudomonas (MEZ, TIC), ≤ 16 for other gram negatives

[b] ≤ 64 for Pseudomonas; ≤ 16 for other gram negatives

[c] ≤ 8 for netilmicin

[d] ≤ 2 for ofloxacin

* very unusual to encounter an isolate susceptible to this drug

See Gilardi reference also

APPENDIX 2C

THESE ANTIBIOGRAMS SHOULD SERVE AS GUIDELINES ONLY - EXCEPTIONS WILL OCCUR.

Gram Positives Bacteria Antibiograms

ORGANISM	AMP	AMP/ SULB	PEN	OXA	1ST CEFS	CHLOR	CLIND	ERY	GENT	TETRA	TMP/ SMZ	VAN
MIC Breakpoint for Susceptibility (μg/mcl)	see below	≤ 8	see below	≤ 2	≤ 8	≤ 8	≤ 4	≤ 4	≤ 4	≤ 4	≤ 2/38	≤4
Staphylococcus aureus 1	S	S	S	S	S	S	S	S	S	S	S	S
Staphylococcus aureus 2	R	S	R	S	S	S	S	S	S	S-R	S	S
Staphylococcus aureus 3	R	S	R	S	S	S	R	R	S	S-R	S	S
(Intrinsic OX-R) *Staphylococcus aureus* 4	R	R	R	R	(R)	S-R	R	R	S-R	S-R	S-R	S
(Acquired OX-R) *Staphylococcus aureus* 5	R	(S)	R	R	(R)	S	S	S	S	S	S	S
Coagulase negative Staphylococcus	—variable results—											S
Group D enterococcus	MS		MS	R*	R*	S	R	S-R	R*	R	(R)	MS
Group D non-enterococcus	S		S	S	S	S	S	S	R*	S	-	S
Streptococcus pneumoniae other *Streptococcus* spp.	S		S	S	S	S	S	S	R*	S	S	S
Corynebacterium spp.	variable results except for Corynebacterium JK and D2. JK and D2 are usually resistant to these (some ERY-S and/or TET-S)											S
Listeria monocytogenes	S		S	S	(R)	S	S	S	S	S	S	S

Additional Breakpoints:

	SUSC	MOD SUSC	INT	RES
AMPICILLIN and PENICILLIN				
• enterococci	-	≤ 8	-	> 8
• non-enterococcal, non-pneumococcal streptococci and other Gram-positives	≤ 0.12	0.25-2	-	>2
• Listeria	≤ 2	-	-	>2
AMPICILLIN				
• staphylococci	≤0.25	-	-	>0.25
PENICILLIN				
• pneumococci	≤ 0.06	0.12-1	-	> 1
• staphylococci	≤ 0.12	-	-	> 0.12
VANCOMYCIN				
• enterococci	-	≤4	8-16	>16

* very unusual to encounter an isolate susceptible to this drug.

APPENDIX 2D

THESE ANTIBIOGRAMS SHOULD SERVE AS GUIDELINES ONLY - EXCEPTIONS WILL OCCUR.

Miscellaneous Gram Negatives Bacteria Antibiograms

	B-LAC	AMP	AMOX/ CLAVU	CEFS	ERY	PEN	TETRA	TMP/ SMZ
Branhamella catarrhalis 1	+	R	S	S	S	R	S	S
Branhamella catarrhalis 2	-	S	S	S	S	S	S	S

oxacillin - R; clindamycin - R; aminoglycosides - S

	B-LAC	AMP	AMOX/ CLAVU	CEFS	CHLOR	TMP/ SMZ
Haemophilus influenzae 1	+	R	S	S	S	S
Haemophilus influenzae 2	-	S	S	S	S	S

	CHLOR	ERY	PEN	TETRA
Pasteurella multocida	S	S-R	S	S

clindamycin - R

	AMP	1ST CEFS	CHLOR	CLIND	ERY	PEN	TETRA
Eikenella corrodens	S	S-R	S	R	S	S	S

oxacillin - R; aminoglycosides - R

CEFS = 1st, 2nd, 3rd generation cephalosporins

APPENDIX 3

Conditions Requiring Verification of Antimicrobial Susceptibility Results

Requires Repeat Testing Unless Patient Had Same Isolate From Another Recent Culture:

*Oxacillin-resistant *S. aureus*

*Gentamicin- + tobramycin- + amikacin-resistant gram-negative bacilli

*Penicillin-resistant or relatively-resistant *S. pneumoniae*
(CDC addresses penicillin-resistant isolates only)

*Ampicillin- + chloramphenicol-resistant *Haemophilus influenzae*

Chloramphenicol-resistant *H. influenzae*

Cephalosporin-resistant *H. influenzae*

Penicillin-resistant or moderately susceptible beta-hemolytic *Streptococcus spp.*

Penicillin-resistant or moderately susceptible viridans *Streptococcus spp.* from sterile body sites

Vancomycin-resistant or moderately susceptible gram-positive organisms (except *Lactobacillus spp.*, *Leuconostoc spp.* or *Pediococcus spp.*; moderately susceptible acceptable for *Enterococcus spp.*)

Enterococcus spp. with high level resistance to streptomycin or gentamicin or both

Trimethoprim/sulfamethoxazole-resistant *X. maltophilia*

*patients with these organisms require contact isolation per CDC Infection Control Guidelines

Requires Verification by Repeat Testing or Reexamination of Test or both:

Imipenem-resistant or moderately susceptible gram-negative bacilli (except *X. maltophilia*, *P. cepacia*; *Proteus/Providencia spp.* are less susceptible than other *Enterobacteriaceae)*

Ciprofloxacin-resistant gram-negative bacilli (except *X. maltophilia* or *P. cepacia*)

Isolate where antibiogram is "atypical" for the species

Isolate resistant to all relevant drugs

Isolate where results of related drugs do not correlate

APPENDIX 4

Major Components of a QC Program for Antimicrobial Susceptibility Tests (Projected % of Effort to be Spent on Component Part of the Program as Identified by JH)

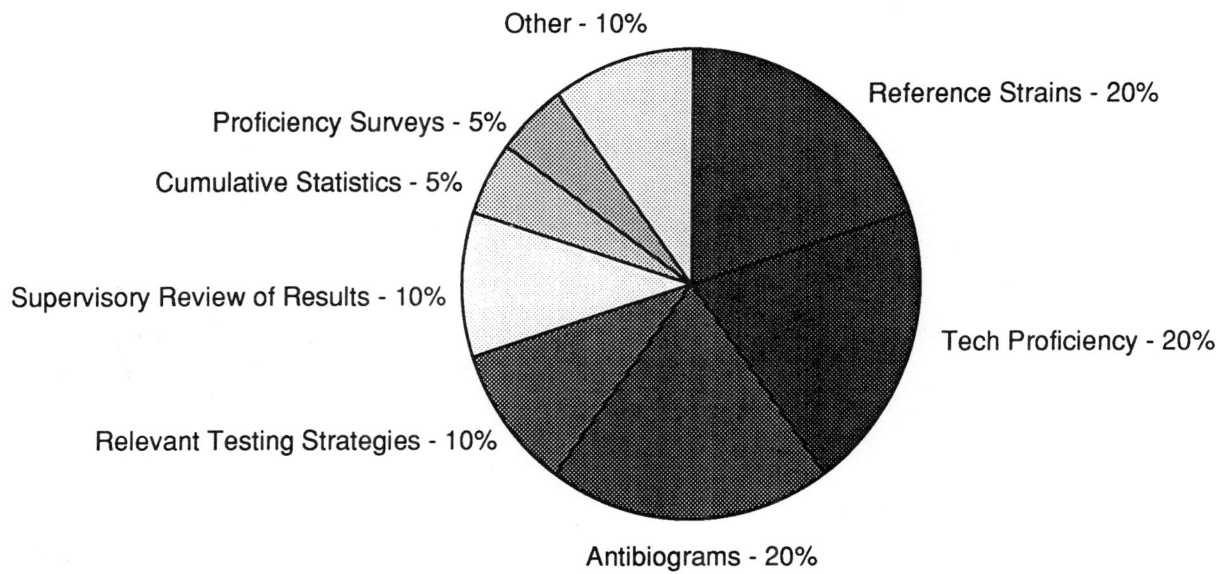

REFERENCES

1. American Type Culture Collection (ATCC). 1989. Catalogue of bacteria and bacteriophages, 17th edition. ATCC, Rockville, MD. (301) 881-2600.

2. A.L. Barry and L.E. Braun, Reader error in determining minimal inhibitory concentrations with microdilution susceptibility test panels. J Clin Microbiol. 13:228-230 (1981).

3. A.L. Barry, M.B. Coyle, C. Thornsberry, E.H. Gerlach, and R.W. Hawkinson, Methods of measuring zones of inhibition with the Bauer-Kirby disk susceptibility test. J Clin Microbiol 10:885-889(1979).

4. A.L. Bisno, WE Dismukes, DT Durack, E.L. Kaplan, A.W. Karchmer, D. Kaye, S.H. Rahimtoola, M.A. Sande, J.P. Sanford, C. Watanakunakorn, and W.W. Wilson, Antimicrobial treatment of infective endocarditis due to *viridans streptococci*, *enterococci*, and *staphylococci*. JAMA 261:1471-1477 (1989).

5. College of American Pathologists (CAP) Proficiency Survey, Bacteriology Survey #D-05. CAP, Skokie, IL. (1991).

6. J. Boyce, J.R. Lonks, A.A. Medeiros, E.F. Papa, and S. Campbell, Spurious oxacillin resistance in *Staphylococcus aureus* because of defective oxacillin disks. J Clin Microbiol. 26:1425-1427 (1988).

7. CDC Guidelines for Isolation Precautions in Hospitals, Nosocomial Infections, Infection Control 4:245-249 (1983).

8. G.L. Gilardi, Identification of glucose-nonfermenting gram-negative rods, Department of Laboratories, North General Hospital, 1919 Madison Avenue, New York, NY. 10035. (1990).

9. J.A. Hindler, Non-traditional approaches to quality control of antimicrobial susceptibility tests, Clin Microbiol Newsletter. 12:65-69 (1990).

10. G.A. Jacoby, and G.L. Archer, New mechanisms of bacterial resistance to antimicrobial agents. NEJM. 324:601-612 (1990).

11. G.A. Jacoby and A.A. Mederios, More extended-spectrum B-lactamases. Antimicrob Agents Chemother. 35:1697-1704 (1991).

12. R.N. Jones, D.C. Edson, and the CAP Microbiology Resource Committee of the College of American Pathologists, Antimicrobial susceptibility testing trends and accuracy in the United States. Arch Pathol Lab Med. 115:429-436 (1991).

13. Medical Letter, The choice of antimicrobial agents, The Medical Letter, 32:41-48 (1990).

14. B.E. Murray, Antibiotic treatment of enterococcal infection, Antimicrob Agents Chemother. 33:1411(1989).

15. B.E. Murray, New aspects of antimicrobial resistance and the resulting therapeutic dilemmas. J Infect Dis. 163:1185-1194 (1991).

16. National Committee for Clinical Laboratory Standards, 1990, Performance standards for antimicrobial disk susceptibility tests - Fourth edition. Approved Standard; M2-A4. NCCLS, Villanova, PA.

17. National Committee for Clinical Laboratory Standards, 1990, Methods for dilution antimicrobial susceptibility tests for bacteria that grow aerobically - Second edition. Approved Standard: M7-A2. NCCLS, Villanova, PA.

18. National Committee for Clinical Laboratory Standards, 1991, Supplemental table for M2-A4, M7-A2, and M11-A2, M100-S3, NCCLS, Villanova, PA.

19. A. Philippon, R. Labia, and G. Jacoby, Extended-spectrum B-lactamases. Antimicrob Agents Chemother. 33:1131-1136 (1989).

20. K.L. Ruoff, Gram-positive vancomycin-resistant clinical isolates. Clin Microbiol Newsletter. 11:1-4 (1989).

21. A. Von Graevenitz, Use of antimicrobial agents as tools in epidemiology, identification, and selection of microorganisms, p. 723-738. In V Lorian (ed.), Antibiotics in laboratory medicine, third edition. Williams & Wilkins, Baltimore (1991).

22. J.A. Washington, II., Current problems in antimicrobial susceptibility testing. Diagn Microbiol Infect Dis. 9:135-138 (1988).

APPLICATIONS OF MEDICAL INFORMATICS IN ANTIBIOTIC THERAPY

R. Scott Evans and Stanley L. Pestotnik

Medical Informatics, Infectious Diseases, and Pharmacy
LDS Hospital and University of Utah School of Medicine
Salt Lake City, Utah

INTRODUCTION

Effective antibiotic therapy is an essential part of hospital care.[1-3] Studies show that patients with rapidly fatal and ultimately fatal underlying diseases have improved chances for survival with appropriate antimicrobial therapy.[4] However, several studies show that there is considerable overuse and misuse of antibiotics.[5-8] Up to two thirds of hospitalized patients who receive antibiotics have no evidence of infection. The major cause for antibiotic overuse results from unnecessary or prolonged surgical prophylaxis. Studies also show that antibiotic prophylaxis should be initiated before the operation to be beneficial.[9,10] Yet, many patients who would benefit from antibiotic prophylaxis do not receive the antibiotics until after the start of the operation.[11,12] The selection of therapeutic antibiotics has been found to be inconsistent with microbiology data.[13,14] This usually results from physicians being unaware of relevant susceptibility results. One study found that the selection of empiric antibiotics was unacceptable 34 percent of the time.[15] The physicians were misinformed about likely pathogens and/or antibiotic susceptibilities. Organisms that were susceptible to antibiotics only a few years ago are now resistant.[16] Antimicrobial resistance is spawned by the overuse and misuse of antibiotics.

The selection of appropriate antibiotics is becoming more difficult due to the introduction of new antibiotics, the change of bacterial pathogens, and antibiotic resistance. Adverse drug events can complicate up to 20 percent of drug therapy in hospitalized patients.[17] In addition, physicians are being pressured to take an active role to reduce the cost of health care. The amount of medical information available to physicians is growing each year. It has been shown that information overload can lead to judgement error.[18] Computerized reminders have been shown to improve physician compliance with predefined care protocols.[19]

Medical Informatics is the rapidly developing scientific field that deals with the storage, retrieval and optimal use of biomedical information, and knowledge for computer-

Antimicrobial Susceptibility Testing, Edited by
J.A. Poupard *et al.*, Plenum Press, New York, 1994

based decision support[20]. Every medical decision is based on information that must be properly recorded and communicated. Medical informatics uses modern computer technology to imporve the availability and use of medical information. This paper describes some of the different ways medical informatics has been used to imporve the use of antibiotics.

Hospital Information Systems

A hospital information system is a computer system that collects and stores patient information in a database. The information can be reviewed by medical personnel through computer terminals located throughout the hospital including the bedside. A knowledge base containing programmed medical logic can be used to monitor the patient information as it is stored and provide computer-based decision support. The system can automatically identify patient situations that need physician or nurse attention. Hospital information systems also provide automatic patient billing and other financial functions.

The HELP (Health Evaluation through Logical Processing) Hospital Information System has been under development at the LDS Hospital and the University of Utah for 20 years.[21,22] This hospital information system is clinically operational at the LDS Hospital in Salt Lake City, Utah. LDS Hospital is a 500 bed, private, tertiary care hospital and a major teaching center for the University of Utah School of Medicine. One of the key features of the HELP system is the integrated database which contains patient information from most clinical areas. The system was designed to provide a medical decision-making capability that could be used to help improve patient care.

Therapeutic Antibiotic Alerts

Microbiology culture and susceptibility results are entered into the laboratory computer by the technologist. Test results entered into the laboratory computer system are automatically sent to the HELP system where they are translated and stored in the patient's computerized medical record. All patient information that is stored in the computerized medical record is screened by a program on the HELP system called the "data driver." The data driver will activate or "drive" certain modules of the knowledge base depending on the type of patient information being stored. If the patient information contains antibiotic susceptibility data, the knowledge base will identify patients who are not receiving antibiotics to which their pathogens are susceptible (Figure 1). Each day a clinical pharmacist contacts the physicians of patients identified by the computer to see if they are aware of the potential problem. During a one year study, the computer system monitored 30,000 microbiology cultures of which 2,157 (7%) contained susceptibility results.[14] The knowledge base generated an "antibiotic alert" for 696 (32%) of the susceptibility results. The physicians were receptive to the pharmacist interaction and willing to discuss the situation. The physicians either changed or initiated antibiotic therapy during the pharmacist contact for 125 of the computer alerts. Another 34 alerts resulted in antibiotic changes within the following 24 hours. The reasons physicians did not change antibiotic therapy for the other computer alerts were: 1) the patient was clinically responding to the prescribed antibiotics, 2) the patient was receiving an antibiotic that was not included on the *in vitro* antibiotic panels, 3) the physician felt the organism(s) isolated from the microbiology culture was either due to contamination or colonization, or 4) the physician decided to use other means of treatment such as surgical debridement or drainage. For 344 (49%) of the antibiotic alerts, the physicians were not aware of the susceptibility results when contacted by the pharmacist.

Prophylactic Antibiotic Timing

We used the HELP system to determine prophylactic antibiotic use patterns and establish baseline postoperative wound rates. During 1985, we found that only 40 percent of patients undergoing surgical operations for which preoperative antibiotics were considered advisable received antibiotics within two hours before the operation.[23] The postoperative wound rate for surgical patients undergoing clean and clean-contaminated operations for which preoperative antibiotics were considered advisable was 1.8 percent.

During 1986, we used the HELP system to generate preoperative antibiotic reminders for patients when antibiotics were advisable. A clinical pharmacist or an AM admissions clerk placed the reminders in the charts of the patients either the afternoon or morning before the operation. The number of patients receiving the first antibiotic dose within 2 hours before the incision rose to 58 percent and the postoperative wound infection rate fell to 0.9 percent ($p < 0.03$). In 1987, we changed the hospital Preoperative/procedure check list so that nurses must indicate whether preoperative antibiotics were given, sent with the patient, or were not applicable. Medical personnel can use a computer program to identify which patients are having operations for which preoperative antibiotics are considered advisable. A follow up study on the timing of prophylactic antibiotics during 1991 showed that 95 percent of patients who had operations for which prophylactic antibiotics were considered advisable received an antibiotic within 2 hours before the surgical incision.

Prophylactic Antibiotic Duration

The hospital information system is also used to monitor the duration of prophylactic antibiotics. Each day the system activates a computer program that monitors every hospitalized patient. The program identifies surgical patients who are receiving prophylactic antibiotics longer than 48 hours after the operation and who do not have evidence of an infection in the computerized medical record. The existence of any of the following information is used by the computer as evidence of a possible infection: 1) a positive or pending microbiology culture; 2) an admission diagnosis of an infection; 3) a Gram stain showing the presence of bacteria or numerous white blood cells; 4) fever (oral > 38° C, rectal > 38.5° C; 5) bacteriuria detected by microscopic examination of urine sediment; 6) an operative infection classification made by the surgeon of either "contaminated" or "dirty;" 7) the patient is in isolation due to infection; 8) a positive chest X-ray showing the presence of pneumonia. In addition, thoracic surgery patients are not identified until all pulmonary artery and arterial catheters are removed. The clinical pharmacists receive a printout of any computer identified patients on their wards. The pharmacists check the identified patients' medical charts and make final judgements whether to stop the antibiotics. An antibiotic "stop order" is placed in the charts of patients determined to be receiving prophylactic antibiotics longer than necessary. During a one year study, the pharmacists decided that seven percent of the identified patients may have had a possible infection.[24] That decision was based on information found in the physicians notes and not contained in the computerized medical record. The computer-generated

```
                    INFECTIOUS DISEASE MONITOR REPORT FOR 28 NOV 1991
                                    FOR PAST 24 HOURS
                                  PRINT TIME: 11/28/91.12:30

***** PATIENT WITH POSSIBLE NOSOCOMIAL PNEUMONIA *****
***** ANTIBIOTIC ALERT *****
***** AZTREONAM WOULD BE THE LEAST EXPENSIVE ANTIBIOTIC *****
 @PAT: 99999999 SMITH, JOHN X.              54   M   ROOM: E111   MR#: 111111
  DOC: 0000 JONES, JOHN X.                 SERVICE: GENERAL SURGERY
  ADMITTED: 11/17/91.02:01            ADMIT DIAG: CHF, CHEST PAIN, R/O MI
  PREV. ADMIT  10/12/1991      PREV. DSCH  11/08/1991
  X-RAY DATA: 11/25/91.04.55  Bacterial Pneumonia
  X-RAY DATA: 11/26/91.08:11  Bacterial Pneumonia
  SURGERY:  Clean                      SURGEON: 999
   11/23/91.08:58   HEART TRANSPLANT - RECIPIENT
  PAT. IS ALLERGIC TO: PENICILLIN
  CURRENT ANTIBIOTICS:
   11/25/91.11:53 TOBRAMYCIN 60 MGM, INJ   Q 12 HRS
   11/25/91/13:28 CEFAZOLIN 1000 MGM, INJ  Q  8 HRS
  CULTURE RESULTS                 -FINAL REPORT-          ROUTINE CULTURE
 SOURCE:  SPUTUM, SUCTIONED                     COLLECTED: 11/25/91.09:35
 STAIN:  NUMEROUS WBCS, FEW GRAM POSITIVE COCCI IN CHAINS,
                   FEW GRAM POSITIVE COCCI IN GROUPS
 RESULT:  PSEUDOMONAS AERUGINOSA   MODERATE GROWTH
                  SUSCEPTIBLE : Aztreonam, Ceftazidime, Ciprofloxacin, Piperacillin,
                   Ticarcillin/Clav,
                  INTERMEDIATE: Cefotaxime, Ceftriaxone, Tobramycin,
                  RESISTANT TO: Amikacin, Amoxicillin/Clav, Cefazolin, Cefoxitin, Cefuroxime,
                   Gentamicin, Imipenem, Tetracycline, Trimethoprim-sulfamethox,
```

Figure 1. Example of a computerized antibiotic alert. The identified patient was not receiving an antibiotic to which a potential pathogen was susceptible. The top three lines are decisions generated by the medical logic contained in the knowledge base. Other pertinent patient information is added to help determine the value of the alert.

stop orders significantly reduced the number of antibiotics being administered beyond 48 hours after operations. The reduction in antibiotic usage resulted in an average savings of $42.00 less per patient and a yearly savings of about $90,000 to patients and the hospital.

Empiric Antibiotic Selection

We statistically analyzed microbiology cultures and antibiotic susceptibility tests during a five year period to identify patient variables that could be used to predict pathogens.[25] We obtained the data from the computerized medical record. The data were electronically transferred to a microcomputer for the analysis. We found that six patient variables could be used to predict what pathogen a patient might have: 1) the site of infection, 2) inpatient vs outpatient, 3) community vs hospital acquired infection, 4) medical service of the patient, 5) sex, and 6) age. Statistical probabilities were then developed to predict the pathogens and select the best antibiotic regimens to treat the predicted pathogens for every combination of the six variables.

Physicians can now use a computer program to help select empiric antibiotics (Figure 2). The program can be accessed from any terminal in the hospital or from physicians' offices and homes. The program uses three levels of logic to determine the most likely empiric antibiotic regimen. The first level is based on the statistical analysis from five years and the most recent six months of microbiology culture and antibiotic susceptibility data. The second level of logic contains rules for the appropriate use of antibiotics according to infectious disease experts. The rules from level two have precedence over the statistical analysis in level one. Level three is based on specific patient information obtained when the program is run. Patient information such as antibiotic allergies, renal condition, the ability to take oral antibiotics, and antibiotic cost is used to make the final antibiotic selection. Each month the microbiology data is automatically updated and a new statistical analysis in level one is performed. The empiric logic provided by the infectious disease experts in level two can be modified or added as required.

An initial study using the empiric antibiotic assistant program was performed on 250 patients with subsequent microbiology cultures and susceptibility results. The program predicted a susceptible and appropriate antibiotic in 238 (95%) of the cases. Further evaluation and testing is currently being performed and analyzed.

Antibiotic Resistance

The computerized knowledge base also monitors all susceptibility results for bacteria that are resistant to specific antibiotics. The most common example is *Staphylococcus aureus* resistance to nafcillin. The computer generates an alert every

LDS HOSPITAL EMPIRIC ANTIBIOTIC ASSISTANT
11/4/91.11:17
99999999 DOE, JOHN D.
SITE = Blood
Inpatient Community-acquired

PAST 5 YEARS ORGANISM	#	(%)	PAST 6 MONTHS ORGANISM	#	(%)
Staph. coagulase neg.	88	(28)	Staph. coagulase neg.	11	(38)
Escherichia coli	81	(26)	Escherichia coli	8	(28)
Strep. pneumoniae	37	(12)	Strep. pneumoniae	4	(14)
Staph. aureus	27	(9)	Strep. viridans	3	(10)
Strep. viridans	27	(9)	Staph. aureus	1	(3)
TOTAL =	260	(84)	TOTAL =	27	(93)

ANTIBIOTIC	(%)	COST/24hr	ANTIBIOTIC	(%)	COST/24hr
Cefotaxime	(87)	$ 27.45	Vanco+Gentamicin	(99)	$ 27.79
Ceftriaxone	(86)	$ 24.98	Vanco+Aztreonam	(99)	$ 51.14
Cefuroxime	(85)	$ 35.46	Vanco+Cefotaxime	(96)	$ 46.85
Cefazolin	(83)	$ 12.00	Vanco+Cefuroxime	(96)	$ 54.86
Vanco+Gentamicin	(99)	$ 27.79	Vanco+Ceftriaxone	(96)	$ 44.38

SUGGESTED EMPIRIC ANTIBIOTIC: Vanco+Gentamicin
* ANTIBIOTIC ALLERGIES -- Sulfonamides
** TOXICITY PROBLEMS -- None identified

Figure 2. Example of the computer display for the patient specific empiric antibiotic assistant program. The most likely pathogens and antibiotic regimens can be compared based on the data from the past five years and the most recent six months.

time a specific bacteria is resistant to specified antibiotics. The computerized antibiogran on the HELP system allows the infectious disease department, the microbiology laboratory and physicians to monitor changes in antibiotic resistance patterns (Figure 3). Th susceptibility data in the antibiogram is automatically updated each month. The program car be run from any computer terminal located in the hospital. The user chooses a specifi bacteria and the susceptibility results of hospital formulary antibiotics is displayed. Th computer sorts the antibiotics by the most susceptible to the least. The computer display als shows the number of times each antibiotic was tested against the specific bacteria and th average cost per 24 hours to treat patients with each antibiotic. The user can compare th susceptibilities from the past five years with the past one year or review the susceptibilitie by nursing division. This program demonstrates that the resistance of some bacteria does var from one nursing division to another.

Many susceptibility reports available from the automated susceptibility systems can b misleading. The reports only show the cumulated results of all of the susceptibility tests tha have been performed by bacteria for a specified period of time. Thus, a few patients wit numerous susceptibility tests for the same pathogens at the same site can bias the results. The logic in the knowledge base on the HELP system examines each susceptibility result an only counts the susceptibility results for the same pathogen from the same infection site fo each patient only once.

LDS HOSPITAL ANTIBIOGRAM
BACTERIA: Escherichia coli
DURING THE PAST 5 YEARS

ANTIBIOTIC	PERCENT SUSCEP.	NUM. TESTED	COST/ 24HR	ANTIBIOTIC	PERCENT SUSCEP.	NUM. TESTED	COST/ 24HR
1.Cefotaxime	100	1882	27.45	11.Cefuroxime	95	2426	35.46
2.Imipenem	100	1795	69.20	12.Cefoxitin	94	2141	34.84
3.Ceftriaxone	100	1629	24.98	13.Ticar/clav	94	1587	41.36
4.Amikacin	99	2489	110.86	14.Amoxi/clav	92	1552	4.08
5.Ceftazidime	99	1882	39.51	15.Trime/sulfa	91	3736	.18
6.Norfloxacin	99	1370	3.48	16.Cefazolin	91	3233	12.00
7.Ciprofloxacin	99	1620	50.26	17.Trimethoprim	91	3238	.36
8.Gentamicin	98	3214	8.39	18.Piperacillin	72	3131	47.12
9.Tobramycin	98	2421	27.75	19.Sulfamethoxaz	69	3245	8.00
10.Aztreonam	98	1549	31.74	20.Ampicillin	68	3968	1.72

Figure 3. Example of the computerized antibiogram at LDS Hospital. Antibiotics are sorted by the most susceptible to the least. The microbiology data used by the antibiogram is automatically updated and reanalyzed each month.

Adverse Drug Events

The Joint Commission on Accreditation of Healthcare Organizations (JCAHO) requires hospitals to identify and report adverse drug events.[26] During a one year period, voluntary reporting of adverse drug events (ADEs) resulted in nine ADEs being identified. Adverse drug events were reported by nurses filling out incidence reports and turning them in to the Quality Management Department.

A computer program was developed on the HELP System and allows nurses, physicians, or pharmacists to report patient symptoms that can be the result of an ADE.[27,28] Pharmacists can use the program to report a sudden request to change or lower the dosage of drug therapy. This program is also available at any terminal in the hospital and is included in the computerized nurse bedside charting. A knowledge base monitors and analyzes all information that is entered from the ADE reporting program. The knowledge base also monitors all drug orders, certain laboratory test results, and laboratory drug level results for indications of possible adverse drug reactions. Each day a computer program prints a list of all patients who have been identified as having possible adverse drug events during the previous 24 hours (Figure 4). The possible adverse drug events identified by the computer are verified by either a clinical pharmacist or a research nurse who use a computerized verification program.

```
              POSSIBLE ADVERSE DRUG EVENT REPORT FOR 27 NOV 1991
                              FOR 1 DAY BACK
                         PRINT TIME: 11/27/91.8:30

***** 11/26/91.22:40  GENTAMICIN TROUGH > 2 *****
@PAT 00000000 SMITH, SAM S.              58   M    W636            MR#: 111111
 DOC: 1111 SMITH, SALLY S.
 ADMITTED: 11/14/91.21:43                ADMIT DIAG: FEVER, S/P LIVER TRANSPLANT
 NO PREV. ADMIT
 CURRENT DRUGS
  11/18/91.10:30 LORAZEPAM 1.0 MGM, TAB   PRN
  11/18/91.18:15 GENTAMICIN 110.0 MGM, INJ   Q 12 HRS
  11/18/91.18:15 NORMAL SALINE 100 ML, INJ   Q 12 HRS
  11/19/91.15:02 IBUPROFEN 600 MGM, TAB   PRN
  11/19/91.15:04 MORPHINE 2.0 MGM, INJ   PRN
  11/19/91.17:36 CYCLOSPORINE 300 MGM, CAP   Q12 HRS
  11/19/91.17:36 PREDNISONE 15.0 MGM, TAB   Q 12 HRS
  11/21/91.14:24 AMPHOTERICIN B 25.0 MGM, INJ   Q 24 HRS
  11/21/91.14:24 ACETAMINOPHEN 650 MGM, TAB   AS DIRECTED
  11/21/91.14:25 DIPHENHYDRAMINE 25 MGM, CAP  AS DIRECTED
  11/22/91.15:15 VANCOMYCIN 1000 MGM, INJ   Q 24 HRS
  11/22/91.15:21 PENTOXIFYLLINE 400 MGM, TAB   Q 6 HRS
 DISCONTINUED DRUGS
  11/18/91.10:54 VANCOMYCIN 1250 MGM, INJ   Q 24 HRS
  11/19/91.08:35 MEPERIDINE 10.0 MGM, INJ
  11/19/91.08:35 MIDAZOLAM 3.5 MGM, INJ
```

Figure 4. Example of a computerized adverse drug event alert. The patient was identified because of a gentamicin blood level of 2.2.

The computerized adverse drug reaction monitor helped to identify 401 verified adverse drug reactions the first year of use and 598 during the second year. Antibiotics accounted for almost a quarter of the adverse drug events with analgesics being the only drugs that caused more adverse events. Physicians are now notified by either the clinical pharmacist or research nurse when their patients' have verified adverse drug reactions.

SUMMARY

The Infectious Disease Society of America is concerned about the excessive and inappropriate use of antibiotics in U.S. hospitals.[29] Applications of Medical Informatics can help improve the use of antibiotics and help improve patient care by monitoring and managing enormous amounts of patient information.

Monitoring the duration of every antibiotic ordered in the hospital or keeping tract of the antibiotic susceptibilities for five years are examples of tasks better performed by computers. The impact of computers in medicine is seen by some as disappointing. The computer revolution has not had the impact in medicine experienced by other areas. The acceptance and use of computers by medicine will be evolutionary rather than revolutionary. In 1979, the MYCIN project demonstrated that the computer could aid physicians in the selection of antibiotics.[30] However, MYCIN was never clinically used because physicians were required to enter all patient information into the computer. The development of computerized medical records is an essential step to further the development and implementation of computer-aided decision support.

The science of Medical Informatics is still relatively new but is emerging as a distinct academic field.[31] A few hospitals are now installing information systems and have determined that these systems will play an essential role in their ability to survive into the next century. The telephone and the automobile have been recognized as two of the most important tools for improving medical care during the past 100 years. People could more readily get medical care and the time to transmit medical information was greatly reduced through physician use of the telephone and automobile. The computer is a tool that can be used to help physicians manage the great amount of medical information being generated every day. The computer can also alert the physician of patient conditions that need attention. However, it is the physician who must use and apply the computer provided information. Thus, the computer will assist but not replace physicians in providing medical care.

ACKNOWLEDGEMENTS

This work was supported in part by Grant HS 06028 from the Agency for Health Care Policy & Research.

REFERENCES

1. C.M. Kunin, T. Tupasi, W.A. Craig, Use of antibiotics: A brief exposition of the problem and some tentative solutions, Ann Intern Med 79:555 (1973).
2. J.E. McGowan, M. Finland, Infection and antibiotic usage at Boston City Hospital: Changes in prevalence during the decade 1964-73, J. Infect Dis 129:421 (1974).
3. H.E. Simmons, P.D. Stolley, This is medical progress? Trends and consequences of antibiotic use in the United States, JAMA 227:1023 (1974).
4. C.S. Bryan, K.L. Reynolds, E.R. Brenner, Analysis of 1,186 episodes of gram-negative bacteremia in non-university hospitals: The effects of antimicrobial therapy, Rev Infect Dis 5:629 (1983).
5. D.G. Maki, A.A. Schuna, A study of antimicrobial misuse in a university hospital, Am J Med Sci 275:271 (1978).
6. M. Shapiro, T.R. Townsend, B. Rosner, E.H. Kass, Use of antimicrobial drugs in general hospitals: patterns of prophylaxis N Engl J Med 301:351 (1979).

7. D.A. Leigh, Antimicrobial usage in forty-three hospitals in England, J Antimicrob Chemother 9:75 (1982).
8. C.M. Kunin, S.T. Chambers, Responsibility of the infectious disease community for optimal use of antibiotics: views of the membership of the Infectious Diseases Society of America, Rev Infect Dis 7:547 (1985).
9. J.F. Burke, The effective period of preventive antibiotic action in experimental incisions and dermal lesions, Surgery 50:151 (1961).
10. D.W. Burdon, Principles of antimicrobial prophylaxis, World J Surg 6:262 (1982).
11. G.J. Jogerst, S.E. Dippe, Antibiotic use among medical specialties in a community hospital, JAMA 245:842 (1981).
12. R.A. Larsen, R.S. Evans, J.P. Burke, S.L. Pestotnik, R.M. Gardner, D.C. Classen, Improved perioperative antibiotic use and reduced surgical wound infections through use of computer decision analysis, Infect Control Hosp Epidemiol 10:316 (1989).
13. W.H. Cannon, D.C. Hale, J.M. Matsen, Program Abstract. 20th Intersci Conf Antimicrob Agents Chemother, New Orleans, LA. abstr. no. 396 (1980).
14. S.L. Pestotnik, R.S. Evans, J.P. Burke, R.M. Gardner, D.C. Classen, Therapeutic antibiotic monitoring: Surveillance using a computerized expert system. Am J Med 88:43 (1990).
15. V.L. Yu, G.P Stoehr, R.C. Starling, J.E. Shogan, Empiric antibiotic selection by physicians: Evaluation of reasoning strategies, Am J Med Sci 301:165 (1991).
16. J.E. McGowan, Improving antibiotic use has become essential - Can surgery lead the way? Infect Control Hosp Epidemiol 11:575 (1990).
17. J. Porter, H. Jick, Drug-related deaths among medical inpatients, JAMA 237:879 (1977).
18. B.L. Drinkwater, Performance of civil aviation pilots under conditions of sensory input overload, Aerosp Med 38:164 (1967).
19. C. J. McDonald, Protocol-based computer reminders, the quality of care and the non-perfectibility of man, N Engl J Med 295:1351 (1976).
20. E.H. Shortliffe, L.E. Perreault, G. Wiederhold, L.M. Fagan, Medical Informatics: Computer applications in health care, Addison-Wesley, New York (1990).
21. T.A. Pryor, R.M. Gardner, P.D. Clayton, H.R. Warner, The HELP System, J Med Sys 7:87-102 (1985).
22. T.A. Pryor, The HELP medical record system, MD Computing 5:22-33 (1988).
23. R.A. Larson, R.S. Evans, J.P. Burke, S.L. Pestotnik, R.M. Gardner, D.C. Classen, Improved perioperative antibiotic use and reduced surgical wound infections through use of computer decision analysis, Infect Control Hosp Epidemiol 10:316 (1989).
24. R.S. Evans, S.L. Pestotnik, J.P. Burke R.M. Gardner, R.A. Larsen, D.C. Classen, Reducing the duration of prophylactic antibiotic use through computer monitoring of surgical patients, DICP 24:351 (1990).
25. R.S. Evans, J.P. Burke, S.L. Pestotnik, D.C. Classen R.L. Menlove, R.M. Gardner, Prediction of hospital infections and selection of antibiotics using an automated hospital database, Proceedings of the Fourteenth Annual Symposium on Computer Applications in Medical Care. Washington, DC,IEEE Computer Society Press, pp 147-156 (1990).
26. Joint Commission on Accreditation of Healthcare Organizations. Accreditation Manual for Hospitals. Chicago:Joint commission. 1990.
27. R.S. Evans, S.L. Pestotnik, D.C. Classen, S.B. Bass, R.L. Menlove, R.M. Gardner, J.P. Burke, Development of a computerized adverse drug event monitor, Proceedings of the Fifteenth Annual Symposium on Computer Applications in Medical Care. Washington, DC,IEEE Computer Society Press, pp 23-27 (1991).

28. D.C. Classen, S.L. Pestotnik, R.S. Evans, J.P. Burke, Computerized surveillance of adverse drug events in hospital patients, JAMA 266:2847 (1991).
29. J.J. Marr, H.L. Moffet, C.M. Kunin, Guidelines for improving the use of antimicrobial agents in hospitals: A statement by the Infectious Diseases Society of America, J Infect Dis 157:869 (1988).
30. V.L. Yu, L.M. Fagan, S.M. Wraith, W.J. Clancey, A.C. Scott, J. Hannigan, R.L. Blum, B.G. Buchanan, S.N. Cohen, Antimicrobial selection by a computer, JAMA 242:1279 (1979).
31. R.A. Greenes, E.H. Shortliffe, Medical Informatics: An emerging academic discipline and institutional priority, JAMA 263:1114 (1990).

ESTABLISHED ANTIMICROBIAL SUSCEPTIBILITY TESTING METHODS WITH A NEW TWIST - POINTS TO CONSIDER AND A GLIMPSE OF THE FUTURE

Lori R. Walsh

Abington Memorial Hospital
Abington, P A.

INTRODUCTION

Antimicrobial susceptibility testing methods (ASTM) have been in use since 1929 to estimate the effects of antimicrobial agents on microorganisms. Routine results are obtained by placing the antibiotics and organisms together in a medium which will support growth utilizing one of the basic methods: agar diffusion, agar dilution, or broth dilution.

Developments in antimicrobial susceptibility testing methods represent innovative modifications to standard methods. The modifications are based on improved technology and responses to marketing surveys and customer suggestions. The companies which develop microbiology equipment are seeking ways to meet the needs of what is one of the last surviving almost - solely - manual, clinical laboratory sections. To assess the needs of microbiologists, as perceived by manufacturers, I examined three new systems and one system which has recently undergone modification. In this assessment, the qualities of each system should reflect the market demand, and thereby identify the needs of the current microbiology laboratory as perceived by the manufacturers. The results should also provide foresight into the future trends in antimicrobial susceptibility testing.

Four methods will be discussed in this chapter. The methods reviewed are: ALADIN, the newest automated microdilution antimicrobial susceptibility testing method; ALAMAR, a non-automated microtiter system; BIOMIC, a computer program to convert qualitative agar diffusion results to quantitative MICs; and CATHRA, an automated agar dilution system.

BACKGROUND

Antimicrobial susceptibility testing methods range from simple, manual techniques with overnight incubation to highly automated systems which can provide results in as little as four hours. In addition to examining the trends in manufacturing, a survey of microbiologists was undertaken to assess the reasons why each system is chosen and to study the issues facing microbiologists as they made their choice of test systems. In 1989, an international survey of antimicrobial susceptibility testing methods employed in the clinical microbiology laboratory was conducted.[1] The survey asked each microbiologist which system they were using and what they would prefer to use. More than 30 antimicrobial susceptibility testing methods were identified and categorized as belonging to one of four standard antimicrobial susceptibility methods.

The distribution of antimicrobial susceptibility methods is listed in Table 1. The choice of antimicrobial susceptibility methods is dependent on many factors, including environmental limitations, price, microbiologist preference and staffing. The survey was an attempt to establish the number, type and frequency of antimicrobial susceptibility methods used throughout the world, and identify critical issues in the decision process.

Table 1. Distribution of Antimicrobial Susceptibility Testing Methods Employed in the Clinical Microbiology Laboratory Through a Survey of 58 Countries[1]

Method	**# Countries**	**% using method**
Agar Diffusion Methods:	55	95 %
Manual Tube Dilution Methods:	18	31 %
Non-automated Microtiter Dilution Methods:	6	10 %
Automated methods:	20	36 %

The survey identified a many types of agar diffusion tests, most of which incorporate a slight modification or "twist" to either the National Committee on Clinical Laboratory Standards (NCCLS) or WHO methodologies.[2] Other variations represent minor mechanical or procedural changes to NCCLS or ECCLS standard methods.

ISSUES IN ANTIMICROBIAL SUSCEPTIBILITY TESTING

The survey identified the distribution of antimicrobial susceptibility test systems in various countries. Key issues identified by clinical microbiologists in the survey include: the expense of the system, ability to generate revenue for the hospital, ergonomic considerations, the ability to introduce new antibiotics easily, decrease turnaround time, improve laboratory workflow and productivity, and manage data. In the United States, 33 % of clinical laboratories use automated methods. This statistic, which is higher than that found in other countries, can be explained partly by economic ability to pay for the system.

SPACE

The physical size of the laboratory will limit the choice of a testing system. Space is needed for the storage of consumables, as well as the placement of the instrument. Some systems require microtiter panels or other disposables to be refrigerated or frozen. This becomes both an equipment problem, and, in developing or tropical countries, a cost and shipping problem. Delivery of refrigerated or frozen items can be both expensive and difficult in warmer climates. Using a system which requires room temperature storage of disposables may be a solution to this problem.

Automated methods tend to require a larger work area, which may include space for a computer terminal, report printer, inoculator plate or tray reader and incubator. A stable source of electricity, and the availability of a water supply are other factors.

ADDITION OF NEW ANTIBIOTICS

One of the most important issues for some clinical microbiologists is the long delay in the addition of new antibiotics to automated or modified systems. New drugs and the use of new guidelines for existing drugs, are delayed while the manufacturer awaits approval from regulatory agencies, such as the FDA, to modify system specifications or add new drugs.

Pre-manufactured testing panels or trays have fixed configurations of antibiotic dilutions. Manufacturers are reluctant to make the necessary changes to this configuration in order to add a new drug to the panel. This also causes delays in the introduction of new drugs to the test systems. The need to modify computer software to report results on a new drug is also a significant problem for manufacturers. This delay alone may make the more expensive systems (usually those with computerized reporting capabilities) harder to justify, by clinical microbiologists, in the future.

Traditional agar dilution and agar diffusion methods allow the addition of new drugs without the delays identified above.

TURNAROUND TIME

Rapid testing, or the ability to release a susceptibility test result in as little as 2 to 5 hours, has become a "hot", but complicated issue. Although antimicrobial susceptibility reports are generated to provide information to the clinician regarding the choice of antimicrobial therapy for the patient, empiric therapy is usually begun prior to any testing in critically ill patients. The preliminary therapy may be broad spectrum and more expensive, but will continue to be administered until culture and identification results are available.

Recent studies[3] suggest that more rapid reporting of antimicrobial susceptibility testing results in cost savings for the hospital as well as the patient by decreasing the length of stay. Patients can be switched earlier, to a less expensive and more specific or tailored drug. Theoretically, rapid turnaround of results has the potential for decreasing resistance to a multitude of drugs by reducing exposure of pathogenic organisms to a large range of antimicrobics.

There is no doubt that the availability of a result in less time is an asset for any system. However, there are still many problems with rapid tests. Resistance to beta-lactam antibiotics may not be uncovered in systems where there is a short incubation period. In many cases, rapid technology will also translate to expensive technology.

LABORATORY WORKFLOW

An advantage of many automated antimicrobial susceptibility methods is the ability to perform both organism identification and susceptibility tests from the same organism suspension, in many cases, using the same test panel. This will not only decrease the "hands-on" time for each isolate tested, but assure the performance of both the identification and susceptibility on the same sub-population of organism. Automated preparation steps have also improved workflow. "Walkaway" type instruments combine these steps with automatic addition of reagents and "reading" of test trays or panels to further reduce the complexity of laboratory workflow. The volume of testing performed by the laboratory will play a large role in determining the workflow needs.

DATA MANAGEMENT

The ability to provide clear, concise results and to generate additional reports may make a computerized system a necessity for those laboratories which do not have a laboratory information computer system. Originally "computerized" systems could store data and print a susceptibility report. More recently, software allows the storage of data, decision-based analysis, compilation of antibiograms, cost analysis information, individual comparisons of drug effectiveness, and the manipulation of susceptibility data to customize patient reports. If the system can be interfaced to a hospital laboratory computer system, results can be made available to physicians as soon as they are released from the laboratory.

Hospital accrediting agencies are increasingly interested in the ability to use laboratory reports to improve quality control, quality assurance and accuracy of printed material. A data management system could help the laboratory director keep up with new regulations and guidelines.

ESTABLISHED METHODOLOGIES WITH A NEW "TWIST"

Many of the above issues have been addressed by the manufacturers of the four systems presented below. Each system represents a modification of a standard procedure. BIOMIC, ALAMAR and ALADIN are the newest methods. CATHRA is included as an example of a "twist" to one of the more classic methods, agar dilution:

BIOMIC	Giles Scientific	Disk Diffusion
CATHRA	Automed	Agar Dilution
ALAMAR	Alamar	Non-Automated Microdilution
ALADIN	Analytab Products Sherwood Medical	Automated Microdilution

THE BIOMIC SYSTEM

BIOMIC was first introduced as the BIOGRAM system in 1985 as a means to expand the capabilities of the disk diffusion method. It is a software program that can be installed on any IBM-compatible computer with 510K RAM and DOS version greater than 2.0 (See Fig. 1). The basic software package calculates, screens and prints qualitative (susceptible, moderately susceptible-intermediate, and resistant)(S, MS-I, R) and quantitative (minimum inhibitory concentration)(MIC) results. An electronic caliper can be interfaced to the computer. Additional database software allows storage, filing and retrieval of information, data management and epidemiology reports.

The electronic caliper is used to record zone inhibition sizes with the touch of a button, and transfer them to the computer. The qualitative or quantitative interpretation is made using NCCLS guidelines. MIC values are calculated by direct regression analysis using an algorithm built into the system which is read from a continuous scale. Studies by D'Amato, Thornsberry, et. al.,[4] have shown a 96 % correlation with tube dilution MICs.

The software also produces a relative activity graph, Figure 2. which compares activities of the antibiotics printed on the susceptibility report, in various body fluids. This activity is quantitated by an inhibitory quotient which is generated by dividing the attainable drug levels at minimal dosing, by the MIC values. The microbiologist can choose to add this graph to the susceptibility report. Drug cost coding can be added to the susceptibility report to aid in choosing the appropriate, least expensive therapy.

The only limitation to testing a new antibiotic is the availability of a disk. Disks are usually the first susceptibility testing device available once a drug is approved by the FDA. BIOMIC does not provide organism identification. It can be used for testing fastidious organisms and can be used as a compliment to other systems.

BIOMIC system uses the same methodology as Bauer-Kirby, but has made the system easier. The price ($995 - $3084) is higher than the non-automated method, but is still very low when one considers the data management capabilities. BIOMIC data management allows the generation of MIC values, drug cost-per-dose information, and epidemiology reports. The electronic caliper decreases the potential for human error when recording zone sizes.

This system does not employ rapid methodology, it will not produce a result in less than 18 hours. Workflow will be only slightly improved by using the electronic caliper and computer-generated susceptibility report. The system is relatively small in size compared to the instrumentation needed for automated microdilution methods and will utilize less laboratory space. However, to test more than 12 antibiotics at one time, a larger work space will be needed to accommodate the extra agar dilution plates.

Most importantly, BIOMIC software provides laboratories using the established Bauer-Kirby method, a means of generating computerized reports and delivering calculated MIC results.

CATHRA

Agar dilution has been limited in use, mostly to research laboratories or larger facilities that test about 50 organisms a day. Automed's CATHRA system mechanizes the agar dilution method and provides computerized reports and data management. CATHRA is well established as an identification and agar dilution system. It is a good example of the trends in antimicrobial susceptibility testing. The system consists of three components: Repliplate substrates or antimicrobial dilutions, the Replicator inoculator and the Replianalyzer instrument. See Fig. 3.

The principle and methodology of the system is the same as for the standardized agar dilution method. Agar dilution susceptibility plates (Repliplates) are available in a variety of antimicrobial agents and dilutions. The plates are inoculated by the Replicator which consists of an inoculating base with a swing arm assembly to transfer up to thirty-six test and control organisms simultaneously from a seeding tray to the Repliplate. Organism identification is performed by testing growth on a number of Repliplate substrates, and is generated by an algorithm. The Replianalyzer is a computerized data entry system which processes the identification algorithm, furnishes susceptibility reports, statistics and epidemiology reports. It can interface with standard laboratory information systems. An instrument called the Repliscan may be purchased to automate the inoculation of the plates and aid in the reading of the test results. CATHRA provides mechanization to the agar dilution reference methodology. New antibiotics can be incorporated into the system early, only if the laboratory personnel are willing to produce their own agar plates using antibiotic powder. However, some drugs, for example, combination drugs using clavulanic acid are too unstable for this method and are not provided as Repliplates. The full Replianalyzer costs approximately, $12,500 and is less expensive than other automated systems. Computer generated reports and increased data management capabilities may justify the increased cost.

ALAMAR

Many laboratories are now using microdilution systems. The available microdilution systems have much in common, the trays or panels are supplied in dehydrated or frozen form, they can generate MICs or breakpoints, and they can be used with a computer to generate susceptibility and epidemiology reports and in some cases are so automated that they approach truly "walkaway technology". Still there continue to be development of new technology and methodology, as in the ALAMAR and ALADIN systems.

ALAMAR is a non-automated microdilution method. It is unique in that it incorporates calorimetric determination of growth in an MIC microtiter plate methodology. The principle of the system incorporates an oxidation-reduction reaction in which a color change occurs in response to the chemical reduction of the growth medium by the dividing bacteria. The microtiter wells are Blue (in an oxidized state) when the tray is removed from its wrapper, and become Red (in the reduced form) in the presence of growth. This color change occurs throughout the well and eliminates such problems as the interpretation of trailing endpoints. Fourteen to seventeen antibiotics can be tested on one ALAMAR tray.

ALAMAR has further improved the flexibility of the microtiter systems by developing a means of delivering antimicrobial agents to the wells of the microtiter plates so that the configuration of the antimicrobials and their dilutions can be changed with each order if necessary. The flexibility is due to the placement of paper disks in the wells of the tray to deliver both the antibiotic and redox indicator.

In this method, as the organism grows the media undergoes a chemical reduction. Continued growth maintains a reduced environment (Red). Inhibition of growth maintains an oxidized environment (Blue). An MIC result is obtained after 18 - 20 hours incubation at 35°C. The MIC is read as the first well which has undergone no color change as compared to the negative control well. Panels are inoculated and rehydrated by delivering 0.1 ml of 1 x 10^5 CFU/ML bacterial suspension in Mueller Hinton broth into each well with a multi-channel pipettor. In a study by Hadley,et.al.,[5] testing 23 antimicrobial agents, 99.2 % of the results obtained by the ALAMAR system, agreed within one dilution of agar dilution reference method. ALAMAR provides MIC panels in a flexible, easily customizable microdilution format. The system is not automated. It is less expensive than its microdilution method counterparts. ALAMAR makes one universal identification panel for gram + and gram - organisms. Further development will be needed before fastidious organisms can be tested using this system. The panels must be refrigerated, which may take up valuable refrigerator space. There is no instrumentation, so the system takes up less space than automated systems.

ALADIN

The desire for a completely "walk-away" system that can incubate, add reagents to microtiter trays, and read these plates within an enclosed system, has led to the development of new automated microdilution systems. ALADIN, which stands for Automated Laboratory Diagnostic Instrument, is manufactured by Analytab Products, Inc.(API), Sherwood Medical. It is unlike the other walk-away systems in that video-image-processing is the technology used to read the reactions in the microtiter wells.

The VITEK and MICROSCAN walkaway systems can produce results in as little as 2-5 hours. ALADIN is not a rapid system, providing results after an 18-hour incubation. A feature of this system that makes it unique is that it discards the microtiter trays, once the analysis is completed. ALADIN and API's other microtiter system, Uniscept share the same microtiter trays, so that the system can be used without automation if necessary.

The panels are introduced into an incubation chamber on a universal carrier which can accommodate any of the API identification and susceptibility panels. After the appropriate incubation time, reagents are added to the panels, if needed, and the tray and carrier are transported to an illuminated station and aligned in the field-of-view of the video image processor (camera). The examination is performed with the aid of four filters selected by the computer on the basis of analytical requirements previously determined for each test and the need to establish results turbidometrically (for the susceptibility testing) or calorimetrically (for identification).

The analytical requirements are determined by the specific areas of interest in the individual microtubule. The viewing area of the camera contains 262,144 pixels (or 512 x 512 picture elements). The areas of the tray between the wells are masked, and 200 to 300 pixels are used to view each individual well. Each well of a susceptibility panel contains approximately 100 to 150 pixels. The output of the video camera is a specific voltage for each pixel. Each pixel is assigned a gray scale value from 0 to 256 that depicts its intensity of gray. Zero voltage is white, full voltage is black. The image processor determines the number of pixels that exceed a given threshold voltage for each well or area-of-interest within the field of view. The specific results for each well are converted into plus or minus reactions. These results are then used to determine antimicrobial end points. The video imager processes many points of information per well as compared with a single data point computation, as in the more commonly used spectrophotometric readers. API believes this additional information leads to better interpretation of fading endpoints, bacterial aggregates, pellicles, sediments and pinpoint bubbles which might interfere with results in other walkaway systems.

ALADIN has added a new technique to viewing MIC results. Its potential advantage lies in improved interpretation of unusual bacterial growth patterns in microtiter wells. The panel configuration is fairly flexible. API panels can be stored at room temperature. The data management features are a strong selling point, but are available on many other automated systems. The cost is high, $72,000 and the amount of laboratory space utilized is large. The potential improved laboratory workflow may justify this cost, but without the availability of rapid results, the microbiologist may opt for one of the other walkaway systems. Comprehensive evaluations of these systems by a wide range of clinical laboratory settings has not been published at this time.

SUMMARY

The developments seen in these systems allow speculation about future trends in antimicrobial susceptibility testing methods. Microbiology system manufacturers seem to be heeding the call of all industry, for greater automation, enhanced data management capabilities and increased flexibility (see Table 2 below).

Table 2. Systems Trend Summary

Feature	Systems
Automation	CATHRA, BIOMIC, ALADIN
Computerization	CATHRA, BIOMIC, ALADIN
Flexibility	CATHRA, BIOMIC, ALADIN, ALAMAR

Cost seems to be less of an issue. This may be due to the decrease in the availability of medical technologists and the need to find systems with better throughput and increased productivity. Increased automation, data management capabilities, and walkaway technology may justify the additional cost of some of these systems. The computer software package provided with these systems is becoming increasingly important with the focus on quality assurance and utilization. Computer generated data analysis gives the microbiologist the tools to educate physicians through the use of selective reporting functions, antibiograms, cost analysis and drug effectiveness comparisons.

Each of the four systems is unique and will probably find a niche among the various markets that exist in the United States, European and other specialized markets. The lack of automation in the ALAMAR system may be its selling point in those areas where automation is not affordable, but new ways are being sought for ease of interpretation of results. BIOMIC and CATHRA systems may be more beneficial to those microbiologists who do not want to stop doing traditional Bauer-Kirby or agar dilution methods, but require computer enhancements. ALADIN, may fill a niche to which other walkaway systems have not adapted, but because of its expense, will face more demands than the other three systems covered in this review.

REFERENCES

1. Walsh, LR. Thesis: An International Survey of Antimicrobial Susceptibility Testing Methods Employed in the Clinical Microbiology Laboratory, (1990).

2. National Committee for Clinical Laboratory Standards, Methods for dilution of antimicrobial susceptibility tests for bacteria: tests for bacteria that grow aerobically. vol. 5. Publication M7-A. National Committee for Clinical Laboratory Standards, Villanova, Pa. (1985).

3. Trenholme, G.M., R. L. Kaplan, P.H. Karakusis, T. Stine, J. Fuhrer, W. Landau, and S. Levin. 1989. Clinical Impact of Rapid Identification and susceptibility testing of blood culture isolates. J. Clin. Microbiol. 27:1342-1345 (1989).

4. D'Amato, R.D., L. Hochstein, J. Vernaleo, D. Cleri, A. Wallman, S. Gradus and C.Thornsberry, J.Clin. Microbiol. 22:793-798 (1985).

5. Hadley, W.K., D. Yajko, P. Nassos, C. Sanders, S.Jenkins, J. Lewis, P. Gilligan, S. Whittier, J. Carlson, and S. Killian. Multicenter Trial of a New Colorimetric Method for Determining Antibiotic MIC for Gram Negative Bacteria. Abstract, American Society for Microbiology, Washington, D.C. (1990).

MEASURES OF SUSCEPTIBILITY FROM A SPIRAL GRADIENT OF DRUG CONCENTRATIONS

Samuel Schalkowsky

Spiral System Instruments, Inc.
Bethesda, MD

INTRODUCTION

The Spiral Gradient Endpoint (SGE) test utilizes the spiral plating method to deposit a liquid suspension in a spiral pattern on the surface of a pre-poured agar plate. Deposition is in exponentially decreasing amounts as the dispensing stylus moves radially outward, starting from the near-center of the plate. The original, and well established application of this method is for the enumeration of bacteria in the suspension, providing for substantial reduction in time and materials because the variable dilution on one spiral plate serves the same purpose as a multiple number of serially diluted pour-plates; colony counting is done only on the portion of the spiral plate containing well separated colonies [1]. Since its introduction in the mid 1970's, this method has achieved wide-spread use for bacterial enumeration, demonstrating the ability of the instrumentation to maintain accuracy and reproducibility while greatly increasing test efficiency.

For its application to antimicrobial susceptibility testing, a spiral plater (Spiral System Instruments, Bethesda Md.) is used to create a radial concentration gradient of the antimicrobial agent in the agar, decreasing from the near-center of the plate outward, because lesser amounts of the antibiotic-containing stock solution are dispensed as the stylus progresses along the spiral. Test strains can be inoculated on the surface of the agar by swabbing along radial lines or by means of an automated replicator. The Radial Replicator (Spiral System Instruments, Bethesda Md.) utilizes a series of stainless steel capillary tubes to deposit up to 15 radial lines of drops onto the surface of the agar. The above two methods of inoculation are illustrated in Figure 1. The sections of the plates shown in this illustration have been photographed over a template used to measure the radial distance of the growth transition endpoint from the center of the plate. This measurement is then used to compute the concentration of the antimicrobial agent at this location on the plate which, in turn, serves to compute the Minimum Inhibitory Concentration (MIC) of the drug/strain interaction.

Antimicrobial Susceptibility Testing, Edited by
J.A. Poupard *et al.*, Plenum Press, New York, 1994

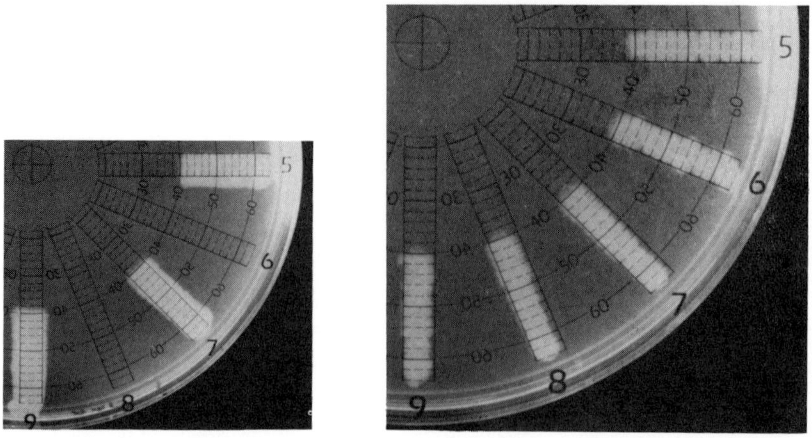

(a) Swabbing (b) Deposition by the Radial Replicator

Figure 1. Alternate Inoculation Methods

The SGE test is *not* a diffusion test method, i.e. it does not rely on drug diffusion to create the gradient. Instead, it utilizes the precisely known volumetric deposition rates, and the known concentration of antibiotic in the stock solution, to determine the amount of antibiotic deposited on the surface at any location on the plate. However, diffusion will alter the deposited gradient. A correction for diffusion is therefore incorporated into the SGE methodology, performed by the SGE software, based on the molecular weight of the antibiotic, radial location on the plate and the agar height. A distinction is also made in the correction algorithm between aerobes and anaerobes since diffusion has a more pronounced effect when testing the slower growing anaerobes.

Evaluations utilizing aerobic test strains showed acceptable correlation to results from parallel tests with the standard agar dilution method even without correction for diffusion [2,3]. Including the diffusion correction algorithm, evaluations of anaerobic test strains by Hill et al.[4], and Wexler et al.[5], utilizing 1,079 and 1,238 on-scale comparisons, respectively, showed 90 percent of the SGE results to be within one dilution of the standard agar dilution MIC determinations.

Fifteen cm pre-poured agar plates are used to provide a concentration range on one SGE plate equivalent to up to 8 twofold dilution plates in the standard agar dilution test. Since the SGE test is performed with pre-poured agar plates, it is significantly less labor intensive and also more convenient than the standard agar dilution method. It also became apparent from the studies on anaerobes as well as aerobes that the SGE test provides significant improvements in reproducibility and sensitivity, attributable to the continuity of the drug concentration gradient from which endpoint measurements are made.

The principal performance measure of existing test methods, the MIC, is a loosely defined parameter and allows a wide range of acceptable variation (+/- one twofold dilution). While this may be adequate for twofold dilution tests, a better defined, and more precise descriptor of antimicrobial action is needed for the continuous SGE method. Such a definition of the MIC, as a discrete rather than twofold value, has been developed and will be described herein. Although this more explicit definition of the MIC is essential for more effective utilization of the SGE method, it is also applicable to the interpretation of standard (twofold) test results, which will also be discussed.

The ability to define a discrete endpoint, and to measure it with greater sensitivity, provides a means for more accurate quantitation of the efficacy of drug combinations. Recommended procedures for synergy/antagonism testing by the SGE method will be described.

Although not a gradient test, use of the spiral plating method to create two nested spirals will be described, illustrating its potential use for assessing drug inactivation due to high levels of beta-lactamase production.

MINIMUM INHIBITORY CONCENTRATIONS

Endpoint Uncertainties

Problems with reading endpoints in standard agar dilution tests have previously been known[6,7], but have become more explicit in the SGE test because of the greater detail shown by the continuous gradient in the growth transition region. While relatively sharp transitions, illustrated in Figure 1, are more common with aerobes, growth tailing, extending over a range of one or more twofold dilutions, are frequent with anaerobes. Examples of such tailing are shown in Figure 2. The most frequently observed growth transition, with anaerobes as well as aerobes, is the discrete colony tail shown in the lowest part of the growth in Figure 2(a). Since only one endpoint can define an MIC concentration, it is not obvious whether the beginning of this tail or its

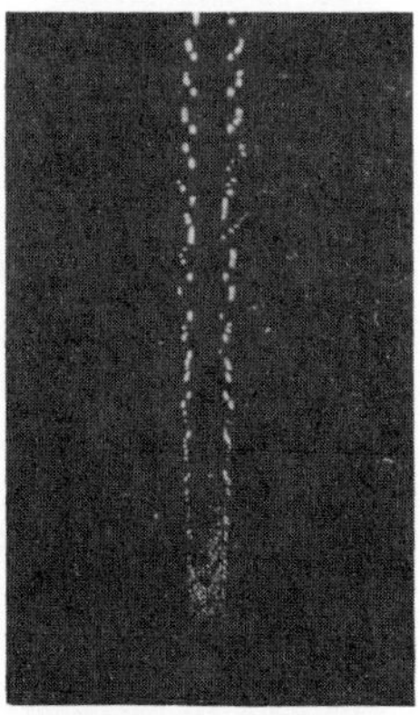

(a) *B. fragilis* group organism vs cefoxitin (courtesy of Gale Hill, Duke Univ.)

(b) *B. fragilis* group organism vs ceftizoxime (courtesy of Hannah Wexler, VA Wadsworth Med. Center)

Figure 2. Growth Transition Tails

end should be used as the basis for computing an equivalent twofold value for comparison with the standard agar dilution result. The discrete MIC will provide a basis for the above computation. The tail, or haze illustrated in Figure 2(b) can extend over a range equivalent to a number of twofold dilutions and also raises the question of endpoint selection for MIC computation purposes. The discrete MIC will facilitate interpretation of the various growth transition features commonly encountered.

Discrete MIC Definition

The definition of a Discrete MIC (DMIC) is made in the context of the time-kill curves shown in Figure 3(a)[8]. Linear kill curves are expected to persist for an initial period of time, provided the drug is applied at the start of the log phase and exposure time is limited to a few generations. In addition, the initial inoculum concentration must be low in order to avoid the presence of more resistant, but less frequently occurring cells of the test population.

In Fig. 3(a), positive slopes, represented by angles measured counter-clockwise from the horizontal exposure-time axis, define restrained growth curves. Negative slopes define population reduction (kill) curves.

Fig. 3(b) will be referred to as an antimicrobial activity plot. It is obtained, theoretically, by producing a large number of initial kill curves and plotting their slopes against the drug concentrations which produced them. The activity plot is therefore a transformation of the information obtained from kill curve plots.

At very small drug concentrations, the slope of the activity plot remains the same and is equal to the slope of the control curve. The first *observable* change in slope occurs when the drug concentration is increased to a value defined as the Minimum Activity Concentration (MAC). At this concentration there would be a measurable reduction in the slope of the kill curve relative to the slope of the control curve.

As drug concentrations are increased beyond the MAC, the positive slope (in the restrained growth region) will continue to decrease until a zero slope concentration is reached. The Discrete MIC (DMIC) is defined as the drug concentration at which the slope of the kill curve is zero. At the DMIC the initial time-kill curve is horizontal, coinciding with the exposure time axis, and the activity curve crosses the drug concentration axis. Thus, at the DMIC, the size of the initial test population remains essentially unchanged. From an observation point-of-view, the DMIC is therefore defined as the drug concentration showing no change relative to the initial (drug-free) inoculum population size.

Figure 3(c) shows the relationship between the tail of the SGE streak and the activity plot, which, in turn is relatable to the kill curve plot. Thus, Figure 3 provides a means for relating observations in the SGE (or standard) agar tests to broth dilution tests.

In Fig. 3(c), TBR is the Tail Beginning Radius, which defines the drug concentration where growth ceases to be confluent. This corresponds to the Minimum

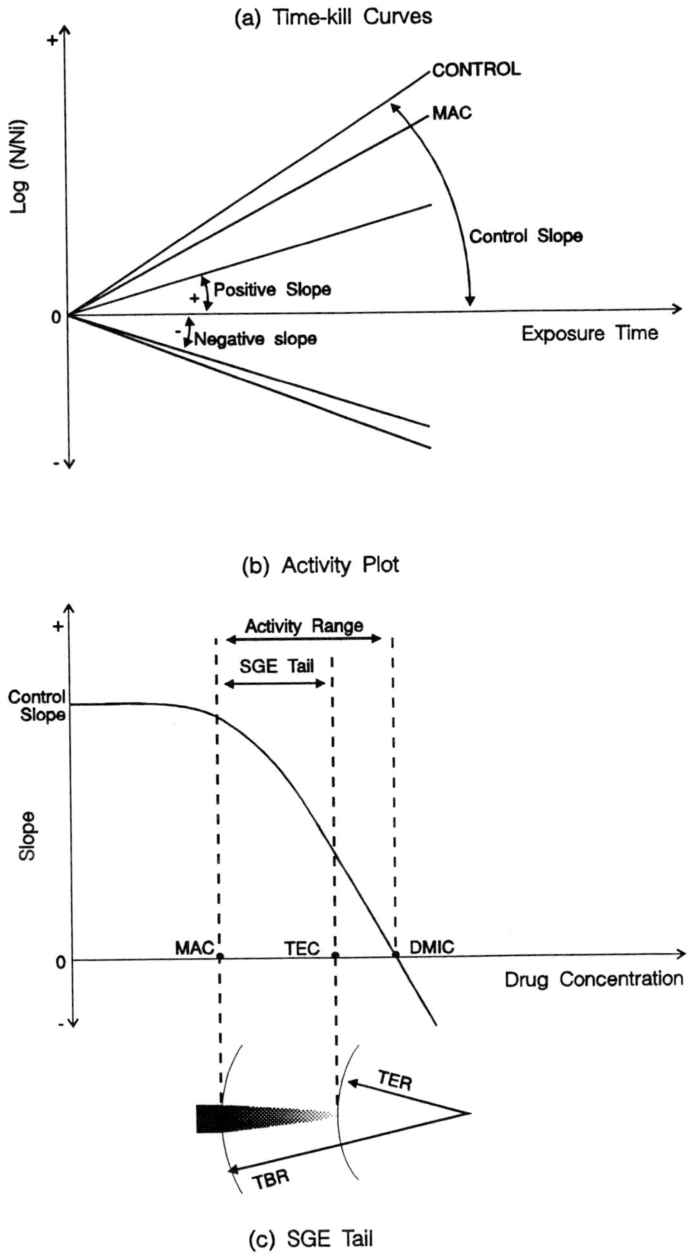

Figure 3. Definition of Discrete Minimum Inhibitory Concentration

Activity Concentration defined for kill curves as the first observable change in slope. Because the MAC is defined by the observation of a test, and can therefore be different in agar and broth tests, it is not used to define the MIC. It is, however, the most readily measurable concentration in the SGE test.

Referring to Fig. 3(c), TER is the radial location where the tail of the SGE streak ends. TEC in Fig. 3(b) is the corresponding Tail Ending Concentration of the drug. It is sensitive to the size of the inoculum and is more variable than the MAC. Furthermore, the TEC is not easily relatable to the DMIC in a quantitative manner. However, it does tell us that the DMIC must be beyond it, at a higher drug concentration, where there are no colonies present, because the initial number of deposited cells remains unchanged.

The DMIC is a uniquely measurable quantity in time-kill tests. The measurement could, for example, consist of obtaining a number of initial kill curves at drug concentrations above and below the DMIC. The DMIC could then be computed from a linear regression of their slopes, or obtained graphically from a plot of the data. This procedure is, clearly, not suitable for routine testing. However, it can serve to resolve uncertainties for classes of drug/strain observations, e.g. whether a hazy part of an SGE streak, or its equivalent standard agar dilution observation, should be considered to be outside or within the MIC value. This procedure has been explored by Summanen et al.[9] The principal decision criterion is that for an observed test spot of an agar dilution plate to be within the MIC, the cell population in the spot must be increasing with exposure time to the drug. If the population is decreasing with time, the corresponding spot should be ignored for purposes of MIC determination, even though live cells are present early in the exposure time interval.

The DMIC, as defined above, is a property of the process and not of the test observations. It is therefore equally applicable to broth as well as agar dilution test, even though the former is a function of population change with time while the other is determined by the presence of visible colonies, essentially independent of time - provided sufficient time is allowed for colony growth. The reason for this commonality is that at the DMIC the slope of the kill/growth curve is also independent of time because the size of the population remains unchanged, i.e. as many cells are killed as successfully divide during a division interval; it therefore does not matter how frequently the cells divide.

The Activity Range defined in Fig. 3(b) is the range of drug concentrations between the MAC and DMIC; it describes the rapidity with which the transition from restrained growth to population reduction is achieved. And, as shown in Fig. 3, the growth-transition tail in the SGE test is indicative of the size of the Activity Range. The relevance of the Activity Range derives from the assumption that the restrained growth region, which is observable in the tail of the SGE streak, mirrors, or is indicative of the rapidity with which effective killing will be achieved with increasing drug concentrations in the population reduction region, which is the focus of clinical applications of the antimicrobial agent. A narrow activity range, equivalent to a small tail, is thus the preferred result of the susceptibility test.

The activity plot for a purely bacteriostatic interaction is shown in Figure 4. It would derive from restrained growth curves limited to positive slopes only. In the absence of bactericidal action, the rate of increase of the population is strictly a function of generation time. Since bacteriostatic action implies increasing generation times with increasing drug concentrations, the corresponding kill-curve slopes will decrease. However, the horizontal axis is never crossed and, theoretically, the DMIC is at an infinitely large drug concentration. In practice, a bacteriostatic gradient streak will show an endpoint at measurable drug concentrations because the generation times need only be increased to values which prevent the appearance of visible colonies during the incubation interval. The streak will tend to have a constant width and, at a sufficiently high inoculum concentration, the transition region will appear like a haze. The beginning location of the tail becomes more difficult to locate and it, as well as the ending location of the tail, are more sensitive to incubation temperature and time.

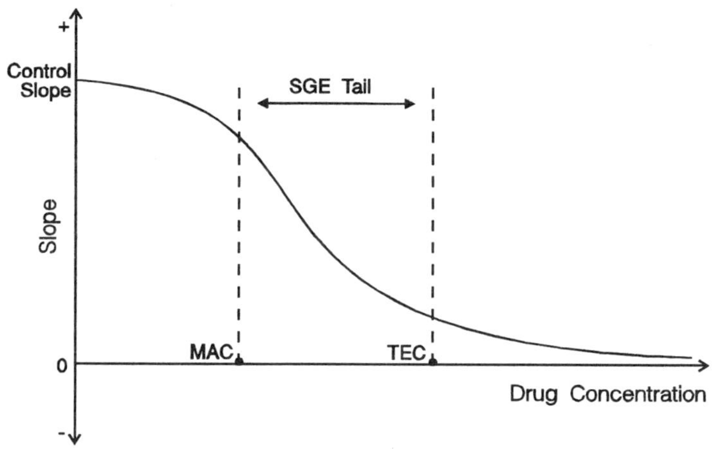

Figure 4. Bacteriostatic Activity Plot

MIC Measures in Twofold Dilution Tests

In the absence of a tailing transition from growth to no-growth, a twofold SGE MIC value, equivalent to that obtained from the standard agar dilution test (or broth dilution test) can readily be obtained by taking the adjacent twofold value above the MAC. Thus, if, based on the tail beginning location, the MAC was found to be, say, 1.65 mcg/ml, the equivalent twofold Gradient MIC - the GMIC, would be 2.0. The corresponding standard twofold MIC will be referred to as the Incremental MIC - the IMIC. The GMIC would, in the absence of a significant tail, be expected to equal the IMIC, since the only possible observations are either full growth or no growth.

The absence of a significant tail is common for the rapidly growing aerobes. This may explain why the NCCLS specification for reading standard agar dilution plates of aerobes requires that the test be repeated if "two or more colonies persist in concentrations of the agent beyond an obvious endpoint"[6], implying that their presence is the result of improper test procedures. But, although uncommon, tailing can be observed, particularly on SGE plates, and their occurrence must be dealt with even for the testing of rapidly growing aerobes.

The presence of tailing growth in the standard agar dilution test would manifest itself as a multiplicity of colonies of approximately equal size. Such tailing can extend to more than one twofold plate, particularly for anaerobes. Since the DMIC is at a drug concentration greater than the end of the tail, the MIC should be based on the highest twofold concentration showing growth tailing. It should not be based on the next higher twofold value, i.e. where no growth is observed, because then the difference between the actual DMIC and the test value could be greater than one twofold dilution.

In the SGE test the length of the growth transition tail is measurable and this measurement can be expressed in twofold dilution units. The GMIC can still be computed from the MAC, because it is less variable than the location of the end of the tail. However, because the DMIC is located beyond the end of the tail, this value is increased by the number of twofold units represented by the tail. For example, if the MAC-based GMIC was found to be 4 mcg/ml and the length of the tail is measured as 1.1 twofold units, the GMIC would be increased to 8 mcg/ml.

Provided growth transition is limited to one twofold dilution, gradient and standard MICs would be the same if based on the above determination protocol. However, a difference of one twofold dilution between the two methods can result if growth transition extends to more than one twofold dilution, because the SGE tail is measured on a continuous, rather than twofold-incremental scale.

As previously noted, the definition of the MIC is based on restrictive test conditions, focusing on an initial, linear kill rate. This rate will not persist if the population contains more resistant cells, not included in the smaller test population used to determine the initial kill rate. The appearance of one or two relatively larger colonies in a no-growth region, or superimposed on tailing growth, should therefore be ignored *for MIC determination purposes*. However, their presence should be noted, as they indicate an undesirable feature of the resistance frequency distribution of the test population, which is likely to be clinically relevant.

The presence of significant growth tailing should also be a part of the test results as it is a measure of the acceleration of bactericidal activity with increasing drug concentrations. In the SGE test, it can be reported as the length of the tail, in twofold dilution units. In the standard agar dilution test it would be measured by the number of twofold plates displaying growth transition. As previously noted, growth tailing is undesirable, as it implies a need to achieve higher drug concentrations for effective bactericidal activity.

Hazy growth transition is likely to be due to largely bacteriostatic activity and should be so noted. Hazy transition should, however, not be confused with observations of an opaque mass in the tail which may be due to the accumulation of dead cells, killed in the course of successive divisions at drug concentrations either immediately above or below the DMIC. A determination as to whether such regions should be ignored, because they are above the DMIC (negative slope), or included as part of the restrained growth region (positive slope), should be made on a class basis by kill curve slope determinations. Summanen et al. made similar determinations by observing viable counts from cells removed from the surface of standard agar dilution plates at increasing times of exposure to the drug[9]. The procedure investigated by Summanen et al.[9], utilizing the change in color of tetrazoleum chloride (TTC) due to contact with live cells, may also be a useful technique for resolving uncertainties between hazy, bacteriostatic growth or the accumulation of dead cells within and beyond the DMIC.

Broth dilution tests also display growth transition in the form of partial turbidity, but these are not explicitly made a part of the MIC determination process. Complete inhibition is specified by the NCCLS for broth dilution tests of aerobes[6]. For anaerobes, complete inhibition is also the basis of MIC determination, but the likelihood of "gradually diminishing amount of growth in successively higher concentrations of the drug" is noted in the NCCLS endpoint criteria, along with the observation that this makes the determination of a sharp endpoint difficult[7]. The advantage of agar dilution is that it is easier to identify and assess the effect of growth tailing.

Interpretive Standards in Disk Diffusion Tests

Breakpoints in disk diffusion tests are obtained from "scattergram" plots of the twofold reference MIC values against the corresponding inhibition zone diameters. The scatter observable in such plots is due, in large measure, to the incremental nature of the twofold MIC values of the reference methods and to the wide range of uncertainty in their measurement[10]. Zone diameters, on the other hand, derive from a continuous scale, with measurement uncertainties depending, at least in part, on the degree and nature of growth tailing. This generally leads to unacceptably low correlation coefficients, if a regression analysis is attempted as the basis for relating reference MIC values to zone diameters.

The zone diameter in disk diffusion corresponds to the Minimum Activity Concentration derived from the tail beginning location on a SGE test. Thus, the zone diameter and MAC are the same discrete points on continuous gradients of drug concentrations. The calibration of zone diameters should therefore be done in terms of the corresponding MAC values in parallel SGE tests. Considerable reduction of scatter should result and acceptable correlation coefficients are likely to be obtained.

In the absence of significant growth tailing, the MAC provides a good estimate of the DMIC and the breakpoint MIC value could therefore be directly translated into its corresponding zone diameter, based on the relationship obtained from the regression analysis. If tailing is present, it is also observable in disk diffusion tests (including the presence of outlier colonies). Thus, although observed on different scales, activity range information in disk diffusion can be related to its counterpart in the SGE test.

MEASURES OF DRUG COMBINATION EFFICACY

Location of the endpoint on an SGE plate is dependent on the strength of the stock concentration of the drug used to produce the gradient. Thus, a higher stock concentration will shift the endpoint, e.g. the Tail Beginning Radius (TBR), toward the outside of the plate, and vice versa. (This feature was used, in part, to validate diffusion correction procedures by testing whether the same MAC is computed regardless of where the TBR is located on the plate.) The ability to shift endpoint locations provides a reliable basis for testing the efficacy of drug combinations, circumventing the fundamental deficiencies of the checkerboard method due to its reliance on twofold dilution methods and its use of a computed null reference - a Fractional Inhibitory Concentration Index of 1 - which cannot be accurately measured in the test.[11]

Figure 5 illustrates the SGE method for evaluating synergy/antagonism of two drugs, each of which has antimicrobial activity defined by its MIC when acting alone. Assume first that there is no growth tailing, resulting in the sharp transitions shown in the illustration. Figure 5.1 shows the different locations of the endpoint of drugs A and B, when tested alone on their separate plates. The next step is to compute a stock concentration - done by the SGE software - which would shift the endpoint of one of the drugs to place it at the same location as that of the other drug. Figure 5.2 illustrates the result if the two drugs were now retested on their own plates. However, this retest does not need to be performed, as it is only necessary to know the stock concentration values which would produce the common endpoint location.

The effect of the combination is tested by producing one plate containing gradients of both drugs, deposited at *half* the stock concentrations which would have placed their individual endpoints at a common location. As shown in Figure 5.3, the result is additive, antagonistic or synergistic if the endpoint remains in the common location, moves inward or moves outward, respectively. In the case of synergy, combinations other than half the individual stock concentrations can be investigated to determine optimum proportions. A quantitative measure of the result, comparable to the FIC Index, can be formulated based on the stock concentrations of the drugs when tested alone and in combination, and the stock concentrations which each drug would require to shift the endpoint by itself to the location of the combination test (computed by the SGE software).

Shifting of the endpoint location need not be the only measure of combination drug efficacy. Thus a beneficial effect may be derived from changes in the nature of growth transition, by removing undesirable tailing - resulting in more rapid killing at relatively lower drug concentrations, and/or by the exclusion of outlier colonies representing undesirable distribution of resistance properties of the population.

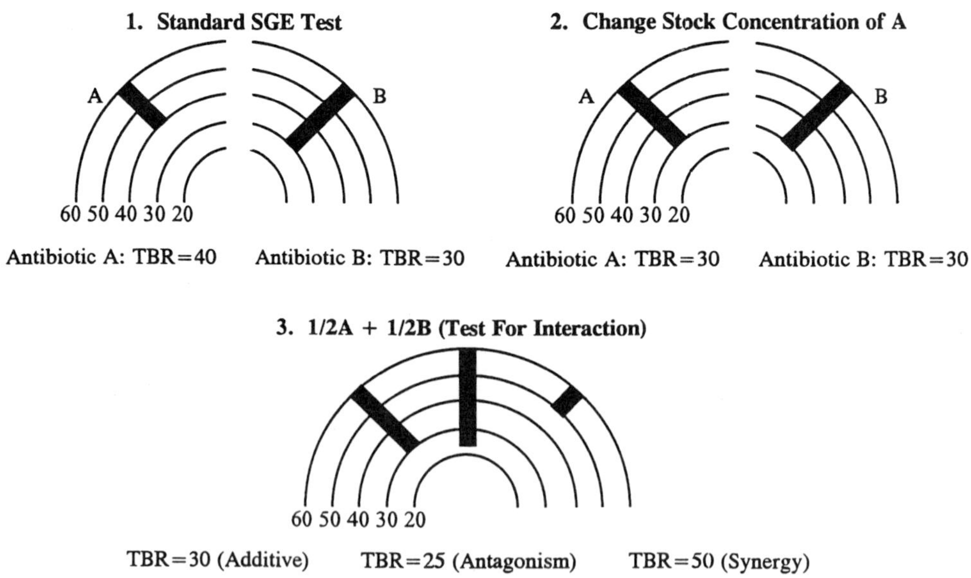

Figure 5. SGE Drug Combination Test

COMPARATIVE TESTING BY THE NESTED SPIRAL METHOD

Although not limited to this application, use of the nested spiral method will be illustrated for its potential utility in assessing the extent of drug inactivation by cell products.

Figure 6(a) shows two nested spirals of a beta-lactamase producing *S. aureus* strain (kindly provided by Kenneth Thomson of Creighton University). The denser spiral was deposited from an undiluted overnight culture, while the other was diluted to a concentration of about 10^8 cfu/ml. Nesting is accomplished by starting deposition of the second spiral 180 degrees from the starting location of the first deposition. Each deposition is at a constant inoculum density, i.e. the same as that sought in preparing a lawn for disk diffusion testing. A variety of antibiotic disks were then placed on the agar surface, to create two nested zones of inhibition around each disk. This method allows convenient visual comparison of the relative effect of the two inoculum concentrations.

Figures 6(b) and (c) are enlargements of the nested zones around two of the antibiotic disks in (a). Although the enhanced growth transition tail shown in (c) suggests a dominant bacteriostatic interaction, further evaluations are needed to provide a better basis for characterizing the inoculum effect of this antibiotic. The shift in the sharp endpoint illustrated in (b) can be more clearly associated with more beta-lactamase production due to the larger inoculum concentration.

Referring to the type of shift illustrated in Figure 6(b), it is likely that the higher concentration of beta-lactamase in the high inoculum spiral will diffuse into the low inoculum region and produces greater drug inactivation in this region than would have otherwise occurred. An *observable* difference in the two zones, therefore, implies a *significant* increase in beta-lactamase production with increased inoculum concentration.

A quantitative assessment of changes in MICs with increased inoculum concentration can be made by an SGE test, with separation between adjacent streaks to preclude diffusion of beta-lactamase between them. This can be used to derive acceptable limits for observable changes in the nested spiral test, which is more convenient for routine testing.

CONCLUSIONS

The SGE test provides a more efficient and more convenient alternative to the standard agar dilution test. In addition, more information is available on the effect of the drug, with superior repeatability and sensitivity.

Definition of the discrete MIC provides a basis for establishing uniform, and mutually consistent criteria for reading the endpoints in twofold agar dilution, SGE and broth dilution tests - for aerobes as well as anaerobes.

The possibility of improved correlation between disk diffusion and MIC breakpoints, using SGE measurements to calibrate zone diameters, needs to be evaluated. It would also be of interest to determine whether this procedure would make the use of disk diffusion feasible for anaerobic bacteria.

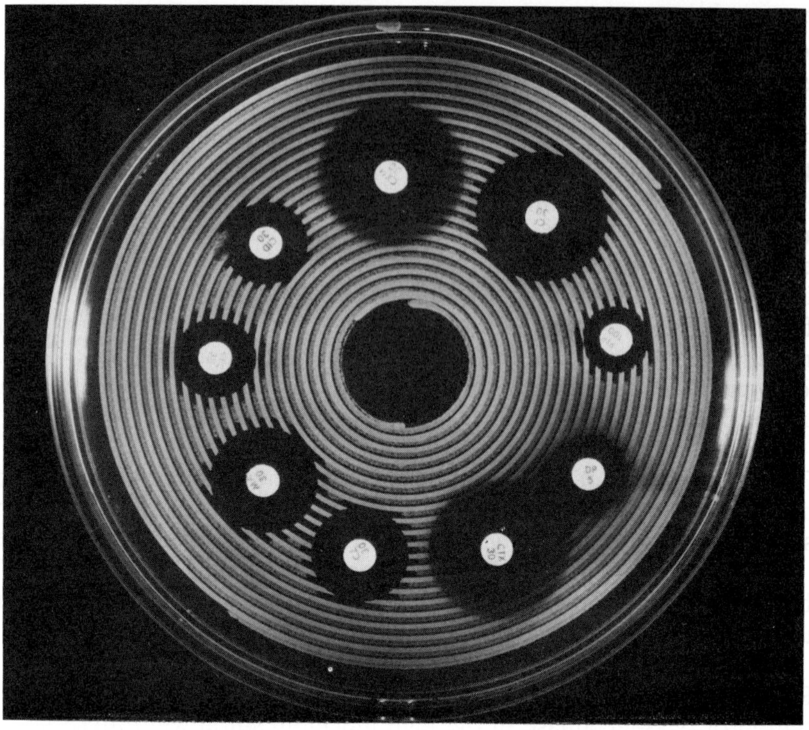

Figure 6(a). Nested Spiral Plate

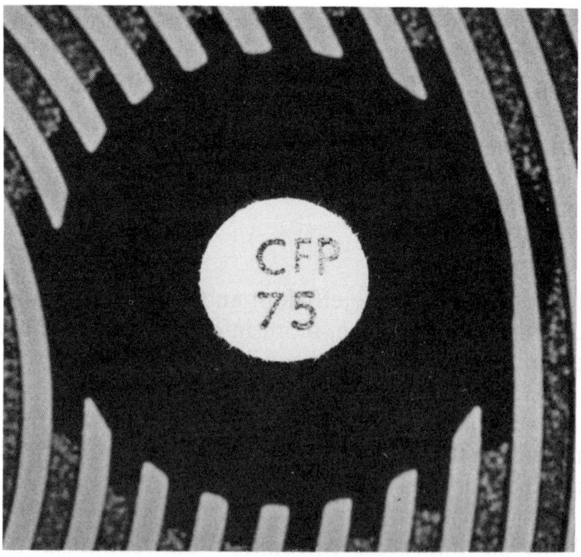

Figure 6(b). Non-tailing Growth Transition

Figure 6(c). Tailing Growth transition

The SGE method provides an unequivocal reference measurement for the quantitative evaluation of drug combination efficacy.

Further studies are needed to assess the utility of the nested spiral method in general, and for the measurement of beta-lactamase production, in particular.

REFERENCES

1. S. Schalkowsky, Plating systems, in: "Foodborne Microorganisms and their Toxins: Developing Methodology," Pierson M. D. and Stern N. J. ed., Marcel Dekker, New York (1986).
2. L. S. Weckbach and J. L. Staneck, Examination of spiral gradient endpoint (SGE) for the determination of MICs [abstract no. C-65]. In: Program and abstracts of the 27th annual meeting of American Society for Microbiology. Washington, DC: American Society for Microbiology, (1987).
3. J. H. Paton, H. A. Holt and M. J. Bywater, Measurement of MICs of antibacterial agents by spiral gradient endpoint compared with conventional dilution method. Int. J. Experimental and Clin. Chemother. 3:1, 31-38 (1990).
4. G. B. Hill and S. Schalkowsky, Development and evaluation of the spiral gradient endpoint method for susceptibility testing of anaerobic gram-negative bacilli Rev.Infect.Diseases 12 suppl. 2, S200-S209 (1990).
5. H. M. Wexler, E. Molitoris, F. Jashnian and S. M. Finegold, Comparison of spiral gradient and conventional agar dilution for susceptibility testing of anaerobic bacteria, Antimicrob. Agents Chemother. 35:6, 1196-1202 (1991).
6. National Committee for Clinical Laboratory Standards, Methods for dilution antimicrobial susceptibility tests for bacteria that grow aerobically, 2nd ed. Approved standard M7-A2. National Committee for Clinical Laboratory Standards, Villanova, Pa. (1990).

7. National Committee for Clinical Laboratory Standards, Methods for antimicrobial susceptibility testing of anaerobic bacteria, 2nd ed. Approved standard M11-A2. National Committee for Clinical Laboratory Standards, Villanova, Pa. (1990).
8. S. Schalkowsky, SGE susceptibility measures. Appendix A in User Guide: Spiral Gradient Endpoint Antimicrobial Susceptibility Test. Spiral System Instruments Inc., Bethesda MD (August 1990).
9. P. Summanen, E. Molitoris, H. M. Wexler and S. M. Finegold, Use of triphenyltetrazolium chloride color reaction as an aid in susceptibility testing of *Bilophila wadsworthia*, Abstract 182, 31st Interscience Conference on Antimicrobial Agents and Chemotherapy (1991).
10. H. M. Ericsson and J. C. Sherris, Antibiotic sensitivity testing. Report of an international collaborative study. Acta Pathol. Microbiol. Scand. (B) Suppl. 217:67-68 (1971).
11. J. A. Korvick and V. L. Yu, Antimicrobial agent therapy for *Pseudomonas aeruginosa*, Antimicrob. Agents and Chemother. 35:11, 2167-2172 (1991).

COMMERCIALIZATION OF NUCLEIC ACID PROBE TECHNOLOGY: CURRENT STATUS

James H. Godsey, Kurt M. Vanden Brink, Luke J. DiMichele, Laura A. Beninsig, W. Richard Peterson, and David G. Sherman

Baxter Diagnostics, Inc.
Microscan Division, West Sacramento, CA

INTRODUCTION

Potential Application of Probes. There are four key diagnostic applications for Nucleic Acid (N.A.) Probe Assays - they are infectious disease diagnosis, genetic disease screening, cancer diagnosis and predisposition to disease screening. Furthermore, identity testing is a non-diagnostic application of N.A. probe assays which includes two specific categories: parenteral identity testing and forensic testing.

What are the benefits of using nucleic acid probe technology in the development of infectious disease assays? Ultimate Level of Specificity - Over the last 25 years, we have seen diagnostic tests evolve from being based upon phenotype characters (colony morphology, staining, growth/no growth, metabolic end products, biochemical tests and antigenic determinants) to being based upon the ultimate level of specificity resident in the genetic code of the organism.

Ultimate Level of Sensitivity - For the last 20 years, innovators of microbiologic diagnostics have dreamed of developing assays sensitive enough to permit direct specimen testing of infectious agents, thus obsoleting general and selective bacteriologic medium after 100 years of service to microbiologists everywhere. With Kary Mullis' invention of the polymerase chain reaction (PCR) 1987[15], the technology which had previously been missing from research laboratories, was suddenly in place. For the first time, it was possible to detect a single bacterial cell or gene copy in a clinical specimen.

How do Probe Assays Work? A nucleic acid probe can be defined as a segment of labeled nucleic acid bases (nucleotides) which hybridize to their complementary bases in the target organism's genome. Probes can be prepared in a variety of ways, including: a)utilizing the entire nucleic acid molecule, such as the 16s rRNA molecule[7], b) creating fragments from 100-1000 bases in length using restriction endonuclease enzymes to cut the chromosome at predefined sites, and c) via a DNA synthesizer.

Antimicrobial Susceptibility Testing, Edited by
J.A. Poupard *et al.*, Plenum Press, New York, 1994

The latter is used to automatically prepare oligonucleotides (oligoprobes) from 12-30 bases in length. Oligoprobes are the preferred reagent for commercial assays because they are more easily prepared, fully characterized and they exhibit faster rates of hybridization than "long" probes. Probes generally have some sort of label attached to them. While the radioactive label, ^{32}P, has been the standard for researchers for a number of years, work over the last 10 years has focused on the development of sensitive nonisotopic labels.

In order to determine whether a probe assay is positive or negative, one must first discriminate between hybridized and unhybridized labeled probes. This is accomplished in one of two ways, a) heterogeneous assay formats and b) homogeneous assay formats. Heterogeneous assays generally contain a second probe, the Capture Probe, which is attached to a solid phase support. The capture probe "captures" the target-signal probe complex and anchors it while unhybridized probe and sample constituents are physically removed from the reaction chamber. Multiple separation steps may be performed to insure that all unhybridized probe has been removed. The target-probe complex is then resuspended and the reaction completed by performing a detection step to determine if label is indeed still present.

There are several types of solid phase supports commonly employed in heterogenous probe assays including: filter paper, microtiter wells, and microparticles (magnetic and non-magnetic). The selection of the solid phase support will have an impact on the final probe assay characteristics. Filter paper (nitrocellulose) and microtiter wells tend to pull target-probe complexes from solution (with the former, the target rather than the capture probe is bound to the solid phase). Microparticle supports, at diameters of ~1um, exhibit near in-solution hybridization time 1/3 - 1/10 that of their microtiter well counterparts.

Homogeneous probe assays are able to discriminate between hybridized and unhybridized label via chemical/physical properties of the hybridization complex rather than the physical separation steps used in the heterogenous formats. While homogeneous assays offer the advantage of assay simplification, they also tend to be somewhat less sensitive than heterogeneous formats.

Detection Schemes for Probe Assays. Labels for nucleic acid probe assay fall into two general groups: Direct Labels and Enzyme Conjugates. Sensitivity of the various labeling strategies varies from approximately 1×10^5 Target Molecules/ml to 1×10^{10} Target Molecules/ml.[20]

Direct labeling strategies offer an advantage over enzyme conjugates in that they: a) eliminate time consuming incubation periods for signal development, b) eliminate 1 or more steps and/or reagents associated with the latter, and c) offer operational specifications which are more robust than enzyme conjugates. Baxter MicroScan has spent the last four years developing and refining the direct label technology referred to as Time Resolved Fluorescence.[3,4]

One important labeling issue to point out is that the sensitivity of the probe system is a result of the combined sensitivities of the labeling strategy and the target amplification technology. If the sensitivity of the label is less than optimal, then the burden for producing the desired combined sensitivity needed to detect an extremely small number of targets, falls completely upon the target amplification technology.

How is the "ultimate level of sensitivity attained" with Nucleic Probe Assays? The invention of the Polymerase Chain Reaction (PCR) is among the most significant discoveries in molecular biology in the last 20 years. It revealed that the N.A. probe research community was trapped in its own "ultimate label" paradigm. That is to say that most of the 1980s were dedicated to developing exquisitely sensitive labels capable of detecting low copy numbers of target from within the complex milieu of a clinical specimen. PCR shifted the focus away from enhancing label sensitivity, to increasing the number of targets available for detection.

Amplification technology can be divided into three categories: a) target amplification, b) probe amplification and c) signal amplification. As implied by its name, target amplification amplifies the sequences of the target genome. PCR was designed to amplify double-stranded DNA and its mode of action is based upon the replication cycle of living organisms. Specificity is gained via the two independent hybridization events of the two primers which are used to delineate the length of target to be amplified, and through the use of a thermostable DNA polymerase (Taq polymerase) to permit the primer extension step to occur under high levels of stringency. The number of targets essentially double with each subsequent PCR cycle, or in a log 2 exponential mode; therefore, twenty cycles of PCR amplification yield a 2^{20} fold level of amplification (approximately one million copies of amplification).

Like PCR, the Transcription Amplification System (TAS) utilizes a two primer approach to insure specificity and to delineate the target region.[12] Unlike PCR, TAS' mode of action is based upon the transcription mechanisms of living organisms. Therefore, at least one of the primers used in each TAS reaction also contains an RNA polymerase binding site. TAS utilizes two enzymes; AMV Reverse Transcriptase to prepare cDNA copies of the target RNA, and T7 RNA polymerase to generate run off transcripts (1,000-10,000 per cycle) of the target region. TAS' ability to initiate amplification from a single stranded RNA molecule enables it to discrimminate between expressed and non-expressed genes. TAS' amplification products are single stranded RNA molecules - ideal for the hybridization/detection step which must follow. 3SR (Self-Sustained -Sequence Replication) is an isothermal version of TAS which allows it to amplify continuously in the absence of thermal cycling.[5,8] Isothermal target amplification is gained by the addition of a third enzyme, RNASE H. RNASE H digests heteroduplexes (RNA-DNA duplexes) in order to free the cDNA strands and initiate the cycle once again. 3SR is capable of producing 1 million to 1 billion fold amplification after just 1 hour of incubation at 42° C. Its isothermal property makes it an attractive technology for automation. Baxter Diagnostics Inc. has an exclusive license to TAS and 3SR from SIBIA, the commercial entity of the Salk Institute, San Diego, CA.

The Ligase Chain Reaction (LCR) was developed by Biotechnica International Inc. and recently sold to Abbott Diagnostics.[2,13,22] LCR utilizes 4 oligonucleotides and a thermostable Ligase enzyme. One pair of probes hybridizes adjacent to each other complementing the entire sequence of bases which represents the target region, while a second pair hybridize to the target sequence in the opposite strand. The thermostable ligase enzyme then enzymatically joins each set of probes to form two double-stranded molecules. The duplex is melted apart and the process repeated. If a single mismatch should occur at the abutting oligonucleotides, no ligation will occur and therefore, no amplification. Like PCR, LCR requires cyclic heating and cooling, with each subsequent cycle resulting in a doubling of the target sequence concentration. Unlike both PCR and 3SR, everything about the LCR target sequence must be known in order for amplification to occur; with PCR and 3SR, only the sequence of the primers used to delineate the length and ends of the target sequence need be known. A potential shortcoming of LCR is that it can only amplify short stretches of nucleic acid (approximately 50 bases). While this may limit its capabilities as a research tool, it probably is not a significant limitation to its diagnostic capabilities.

Q-beta Replicase, discovered by F.R. Kramer, is another example of probe amplification technology.[12,13] An RNA probe sequence is incorporated to a natural template (MDV-1) of the Q-beta replicase. Amplification is initiated when the RNA reporter sequence hybridizes with its complementary sequences in the target DNA. The conformation change, which is brought about by the hybridization event, activates the replicase. The Q-beta replicase can produce 1×10^9 copies of the probe sequence within 30 minutes. This technology is now owned by GeneTrak Systems, Inc.

XMAS TREES is a direct reference to the branched DNA technology of the Chiron Corporation and represents an example of signal amplification technology.[19] Chiron has developed the capability of synthesizing branched DNA molecules (up to 50 branches) that bind to a universal capture probe which has hybridized to the target sequence. Subsequently, a third DNA probe, each of which contains an enzyme conjugate, hybridizes to a universal capture sequence at the terminal end of each DNA "branch". XMAS TREES are therefore capable of attaching approximately 50 enzyme conjugates to a single target sequence. While detection of a single target nucleic acid is not possible with this approach, targets numbering in the range of 1×10^4 to 1×10^5 CFU/ml are detectable.

A universal limitation to the commercialization of amplification technologies is sample cross contamination. Should a single copy of the amplification product contaminate a neighboring negative reaction well, the amplification enzyme(s), nucleotides or buffer mixtures, all subsequent amplification reactions will yield false positive results.

The UNG Procedure is currently available for PCR which provides for the enzymatic degradation of amplification products prior to initiating each subsequent amplification reaction.[14] This process of "chemical sterilization" must eventually be developed for each of the alternative amplification technologies if they are to remain commercially viable.

What key events must occur to deliver infectious disease assays to the marketplace? Diagnostic companies intending to be major players in the N.A. probe assay market for infectious diseases must first acquire and/or develop proprietary positions with each of four core technologies (sample preparation, amplification, hybridization and detection). Failure to secure a proprietary position with one or more of these core technologies would severely limit a company's ability to compete in this marketplace. Secondly, the four technologies must be successfully integrated into a working assay which is compatible with state-of-the-art delivery systems. The first of these delivery systems will be based upon automation in order to make the cumbersome steps associated with the performance of N.A. probe assays transparent to the end user.

How will nucleic acid probe assays be "packaged" for the infectious disease marketplace? A) Manual assays with or without a reader. Gen Probe's Pace2 Assays for *Chlamydia trachomatis* and *N. gonorrhea* represent the premier example in this category and the first commercially successful probe assay. The various manual steps are batched to minimize the actual hands on time per sample and the final detection step performed on a luminometer.

B) Generic Washer/Aspirator/Reader. Baxter Diagnostics, Inc. Bartel's Division has recently introduced its PRIMA (Probe Immunoassay) System designed to take advantage of the many ELISA systems currently found in laboratories throughout the world. The PRIMA System offers both 3SR-based N.A. probe assays and ELISA assays for the same delivery system. The assays use a generic protocol which includes an HRP(horseradish peroxidase) detection scheme.

C) Modified semi-automated Immunochemistry Analyzers. Several diagnostic companies are working to add probe assay capabilities to their existing equipment already present in customers' laboratories. The advantage gained would be the rapid penetration/dissemination of probe assays via the already existing large customer base. The key companies to watch in this category are: Baxter Diagnostic's Stratus System; Abbott Diagnostic's IMX System; Roche Diagnostic's COBAS System; Vitek's Vidas System. The key issue to be addressed by each of the above companies is that without certain modifications and/or add-on modules, the above equipment will not be compatible with probe technology as it has been defined at this writing. Probe assays require incubation temperatures of 100° C for denaturation of double stranded targets and 55°-65° C for hybridization. Immunochemistry analyzers operate at ambient temperatures. Probe assays require dispense volumes of 1-10 ul; immunochemistry analyzers dispense volumes from 100 ul-500 ul. Detection schemes used on today's immuno- chemistry analyzers utilize labeling strategies which tend to be 100-1000 fold less sensitive than some of the more optimal labeling schemes developed specifically for N.A. probe assays. While it's apparent that all of the above issues can be addressed, the mode of implementation will be the key to each company's eventual level of marketing success.

The final category is **Automated analyzers dedicated to nucleic acid probe assays.** Baxter Diagnostics, Inc., Microscan Division has chosen automation as an optimal vehicle for successfully introducing N.A. probe assays to the clinical microbiology community.

MicroScan's goal is to successfully integrated several state-of-the-art proprietary probe technologies including: sample preparation, 3SR amplification, non-magnetic microparticle-based solid phase supports and Time Resolved Fluorescence, into an automated delivery system designed for the N.A. probe marketplace.[3,4]

Time Resolved Fluorescence Concept. Time Resolved Fluorescence was the result of Irwin Wieder's successful attempts[21] to gain access to the ultimate level of sensitivity promised early on by fluorescence technology but never realized previously due in large part to the presence of background fluorescence associated with biologic molecules present in clinical specimens. Wieder was able to show that this background fluorescence was short-lived (on the order of nanoseconds). He hypothesized correctly that if he were to find a fluorescent label whose half life was several orders of magnitude greater than that of the background fluorescence, he would be able to use TIME as the discriminating parameter between target and background fluorescence (Figure 1). Rare earth metals (REM), such as europium and terbium, proved to be ideal candidates for this application due to their long lived fluorescence (1 msec half-life); however, special molecules called chelates, were necessary to capture and transfer the energy required to raise the REM to this long term excited state. Rare earth metal chelates (REMC) can be attached to a nucleic probe using straightforward chemistries. These complexes are quite stable. Probes labeled with MicroScan's REMCs can be read immediately upon hybridization. One advantage of REMCs over other labels, is that the REMC can be re-read any number of times without sacrificing accuracy.

MicroScan Probe Product Development Overview. MicroScan's five step assay will successfully integrate Sample preparation, 3SR amplification, bead-based hybridization and Time Resolved Fluorescence (TRF) detections into a totally integrated assay (Figure 2) and require only 2.5-3 hours to complete, depending on the type of sample being analyzed. A proprietary microtiter tray which possesses wells whose bottom contain a membrane and a liquid waste trap, enables the automated processing of the hybridization/detection steps in the assay.

Evolution of Time Resolved Fluorescence Assay. The impact of automation on N.A. probe processing efficiency is best demonstrated by examining the evolution of MicroScan's TRF assay. There early manual version of the assay required 2.5 hours for one person to complete 16 assays. With the current semi-automated research assay, one person can perform 96 assays in 80 minutes, of which 10 minutes is actual hands on time.

What it the Time Frame for the Introduction of the First Successful Probe Assays for Antibiotic Resistance Genes? While there exists a multitude of N.A. probe assays for infectious disease screening awaiting commercialization, successful strategies for developing and implementing N.A. probe assays for the detection of antibiotic resistance genes are still being formulated and are probably 5-10 years off.

Antibiotic Resistance Gene Nucleic Acid Probe Assays - Technological Hurdles. In order to successfully develop and market N.A. probe assays for the detection of antibiotic resistance genes directly from clinical specimens, one must overcome a number of technological and "political" hurdles, including: In the absence of general and selective bacteriologic culture medium which serve to subdivide unknown isolates into various phenotypically related groups, direct specimen N.A. probe assays must be capable of detecting the presence of approximately 100 different antibiotic resistance genes, as well as identifying ~400 bacterial/fungal species. Even if the technicalities of simultaneously testing such a large number of probes could be worked out, the price of performing large numbers of N.A. probe assays would be cost prohibitive using today's technology.

One of the basic principles of reporting antibiotic susceptibility test results is the requirement to determine the identification of the organism in question. Detecting an aminoglycoside resistance gene present in an enterococci has a much more serious connotation than it would if found in a staphylococcal organism. While Sanchez-Pescador, et.al.[17] were able to demonstrate the detection of taxonomic (*N. gonorrhea*) and antibiotic resistance genes (TEM-1) in the same clinical specimen, the consistent phylogenetic association of the TEM-1 plasmid with penicillin resistance in *N. gonorrhea* significantly limits the applicability of this model system to the dramatically more complicated environment of other clinically significant bacteria. Therefore, much work still remains before N.A. probe assays yield antibiotic resistance data which can be associated at an acceptable level of confidence with the identification of the pathogenic organism in question.

Current N.A. probe assays are designed to detect the presence of a particular sequence and relate that to the presence of the target organism, or, in the case of antibiotic resistance, a target gene. Probes capable of detecting whether a gene is expressed or non-expressed are currently under development but are not widely available. Detection of a non-expressed chromosomal beta-lactamase gene would be the equivalent of a major categorical error (reporting a false resistance) with today's *in vitro* methodologies.

Not all antibiotic resistance mechanisms are clearly associated with a single gene. The detection of intrinsic antibiotic resistance mechanisms may be one of the most difficult challenges for N.A. probe assays. The best example of this phenomena is the ability of an organism to selectively decrease the permeability of its cell membranes to certain antibiotics. The genetic loci controlling this type of antibiotic resistance mechanism may not be fully characterized for several years to come.

Finally, if one makes the assumption that the above hurdles can be successfully overcome, one is then faced with the task of implementation of direct specimen testing in the hospital setting. Technology which produces immediate results for bacterial/viral identification and antibiotic resistance tests performed directly on the clinical specimen would appear to challenge many of the basic principles/practices of empirical therapy. One would expect considerable reluctance by physicians to fully adopt such practices. Furthermore, retraining the medical community on such practices would appear to be a task of major proportions.

Nucleic Acid Probe Assays for the Detection of Antibiotic Resistance Genes - Accomplishments. One can, however, close this discussion on a very positive note; much has been accomplished toward the development of N.A. probe assays for antibiotic resistance genes.

As is always the case, there are pioneers in the clinical microbiology and molecular biology research communities who have looked into the future and realized the potential impact of nucleic acid probe technology on antibiotic resistance testing.[1,6,9,10,16,18] These pioneers have already developed probe assays for approximately fifty of the known antibiotic resistance genes (Table 1) and several of these researchers have begun to work on how best to "package" their assays so as to minimize the task at hand.

In summary, the time frame spanning the decade of the 90s will offer many advances in the area of N.A. probe-based diagnostics for infectious disease. Among the most challenging accomplishments will have been the development of commercial N.A. probe assays for antibiotic resistance genes.

REFERENCES

1. G.L. Archer, E. Pennell, Detection of methicillin resistance in staphylococci by using a DNA probe, Antimicrob. Agents Chemother. 34(9): 1720-1724 (1990).
2. F. Barany, The ligase chain reaction (LCR) in a PCR world, PCR Methods and Applications 1:5-16 (1991).

3. C.E. Bush, L.J. DiMichele, W.R. Peterson, D.G. Sherman, and J.H. Godsey, Solid-phase time-resolved fluorescence detection of HIV PCR amplification product, Analytical Biochemistry, 202 (1):146-151 (1992).
4. C.E. Bush, K.M. Vanden Brink, D.G. Sherman, W.R. Peterson, L.A. Beninsig, and J.H. Godsey, Detection of *Escherichia coli* r RNA using target amplification and time-resolved fluorescence detection, Molecular and Cellular Probes, 5:1064-1079 (1991).
5. C.E. Bush, R.M. Donovan, W.R. Peterson, M.B. Jennings, V. Bolton, D.G. Sherman, K.M. Vanden Brink, L.A. Beninsig, J.H. Godsey, Detection of HIV-1 RNA in plasma from high risk pediatric patients using the self-sustained sequence replication reaction, J. Clin. Microbiol., 30(2):281-286 (1992).
6. P. Coll, K. Phillips, F.C. Tenover, Evaluation of a rapid method of extracting DNA from stool samples for use in hybridization assays, J. Clin. Microbiol. 27(10):2245-2248 (1989).
7. J. DeLey, Intra and Intergeneric Similarities of the Ribosomal RNA Cistrons of Acetobacter and Gluonobacter, Int. J. System. Bact. 30 (1):7-27 (1980).
8. J.C. Guatelli, K.M. Whitfield, D.Y. Kwoh, K.J. Barringer, D.D. Richman, and T.R. Gingeras, Isothermal *In Vitro* Amplification of Nucleic Acids by a Multienzyme Reaction Modeled After Retroviral Replication, Proc. Natl. Acad. Sci. 87:1874-1878 (1990).
9. S. Huovinen, M.L. Klossner, M.L. Katila, P. Houvinen, Plasmid-Mediated Beta-Lactamases among aminoglycoside resitant gram-negative bacilli, Scand. J Infect. Dis. 21 (3):303-309 (1989).
10. G.A. Jacoby, M.J. Blaser, P. Santanam, H. Hachler, F.H. Kayser, R.S. Hare, G.H. Miller, Appearance of Amikacin and Tobramycin Resistance Due to 4'-Aminoglycoside Nucleotidyltransferase [ANT(4')-II] in Gram-Negative Pathogens, Antimicrob. Agents Chemother. 34(12):2381-2386 (1990).
11. D.Y. Kwoh, G.R. Davis, K.M. Whitfield, H.L. Chappelle, L. DiMichele, and T.R. Gingeras, Transciption-Based Amplification System and Detection of Amplified Human Immunodeficiency Cirus Type 1 with a Bead-Based Sandwich Hybridization Format, Proc. Natl. Acad. Sci. 86:1171-1177 (1989).
12. P.M. Lazardi, C.E. Guerra, H. Lomeli, I. Tussie-Luna, F.R. Kramer, Exponential Amplification of Recombinant-RNA Hybridization Probes, Biotechnology, 6:1197-1202 (1989).
13. R. Lewis, Innovative Alternatives to PCR Technology are Proliferating, The Scientist Jan. 21:23-24 (1991).
14. M.C. Longo, M.S. Rerninger, and Hartley, Use of Uracil DNA Glycosylase to Control Carry-over Contamination in Polymerase Chain Reactions, Gene 93:125-128 (1990).
15. K.B. Mullis, and F.A. Faloona, Specific Synthesis of DNA In Vitro Via a Polymerase-Catalyzed Chain Reaction, Methods Enzymol. 155:335-350 (1987).
16. H. Ounissi, E. Derlot, C. Carlier, P. Courvalin, Gene Homogeneity for Aminoglycoside-Modifying Enzymes in Gram-Positive Cocci, Antimicrob. Agents Chemother. 34(11):2164-2168 (1990).
17. R. Sanchez-Pescador, M.S. Stempien, and M.S. Urdea, Rapid Chemiluminescent Nucleic Acid Assays for Detection of TEM-1 Beta-Lactamase-Mediated Penicillin Resistance in *Neisseria gonorrhoeae* and Other Bacteria, J. Clin. Microbiol. 26:1934-1938 (1988).

18. F.C. Tenover, K.L. Phillips, T. Gilbert, P. Lockhart, P.J. O'Hara, J.J. Plorde, Development of a DNA Probe from the Deoxyribonucleotide Sequence of a 3-N-Aminoglycoside Acetyltransferase [AAC(3)-1] Resistance Gene, Antimicrob. Agents Chemother. 33(4):551-559 (1989).
19. M.S. Urdea, J.A. Running, T. Horn, J. Clyne, L. Ku, and B.D. Warner, A Novel Method for the Rapid Detection of Specific Nucleotide Sequences in Crude Biological Samples Without Blotting or Radioactivity: Application to the Analysis of Hepatitis B Virus in Human Serum, Gene 61:253-264 (1987).
20. M.S. Urdea, B.D. Warner, J.A. Running M. Stempien, J. Clyne, and T. Horn, A Comparison of Non-Radioisotopic Hybridization Assay Methods Using Fluorescent, Chemiluminescent and Enzyme-labeled Synthetic Oligodeoxyrobonucleotide Probes, Nucleic Acid Res. 16:4937-4956 (1988).
21. I. Wieder, Method and Apparatus for Improved Analytical Fluorescent Spectroscopy, U.S. Patent 4,058,732 (November 15, 1977).
22. D.Y. Wu, and R.B. Wallace, The Ligation Amplification Reaction (LAR)-Amplification of Specific DNA Sequences Using Sequential Rounds of Template-Dependent Ligation, Genomics 4:560-569 (1989).

IS ONE LABORATORY IN TOWN ENOUGH?

Joel E. Mortensen

St. Christopher's Hospital for Children
Philadelphia, PA

INTRODUCTION

Hospitals have had enormous economic and sociological importance in the United States. The roughly six thousand community hospitals in the United States account for more than 4% of the gross national product. Hospitals directly employ more than three million persons. In addition, hundreds of thousands of people are employed providing goods and services to the hospital industry, along with more than 300,000 physicians.[1,2,4]

Hospitals excel in two quite different areas, finance and technology. Because more than half of many hospitals' revenues come directly from state and federal governments, and because the financial demands of hospitals have grown so disproportionately in a period of tax-cutting and economic constraint, hospitals are increasingly a concern to the public and politicians - hence Diagnosis Related Groups (DRGs). In an almost contradictory arena, the mystical, magical world of high tech medicine is played out. Our society is captivated with the promises of science and technology, regardless of cost, and the hospital remains the focus for some of the most dramatic products of current biomedical research.[4]

The clinical laboratory has been a microcosm of the hospital as a whole, if not a somewhat less visible microcosm. Over the last twenty years, clinical laboratories have undergone a technical revolution. New methodologies such as DNA technologies and Enzyme Immuno Assays along with new equipment have been developed at a dizzying pace. In addition, automation and computerization have become state- of -the-art in clinical laboratories. The net effect of these important developments has been an overall reduction in the cost per test, and in the opinion of most, an increase in the quality and standardization of results.[2] Over the last 15 years, several reports have suggested that the next step in streamlining and improving laboratory services should be centralization or regionalization of laboratory services.[1,2,3] In 1975, Dr. James Prier eloquently discussed this issue from the view of the State Laboratory System. Although his focus was somewhat different, his insights are appropriate to today's questions.[4] I have no intent to fully explore why we want to centralize laboratory services;

however, I will admit that some of the financial wizards' claims can probably be realized. We probably can reduce overhead and management staff, while gaining expertise and savings from volume. The question that many of us have is how can the clinical microbiology laboratory provide state-of-the-art microbiology, offer a full menu of testing, and operate within an ever-shrinking budget? Althougn this question is being asked within the structure of a symposium on antimicriobial susceptibility testing, the issues are the same for other testing methods in microbiology. The largest pieces of the puzzle are information, transportation and education/interaction. Of course, each piece can be further divided.

Specifically, we must be able to move the necessary patient demographic and testing information and the specimen from the site of collection to the off-site processing center to the main laboratory. Then, that information plus the testing results and interpretation must travel to the clinician ordering the test and to the hospital or physician's office of origin. Transportation of specimens in a timely and appropriate fashion can be as difficult as moving the information, and is given to ongoing problems. Finally, what separates the good clinical laboratory from the mediocre is the educational and technical interactions that must occur between all of the players.

I would like to offer some options and solutions to the potential problems associated with the task. Of course, every setting is different according to size, people involved, geography, administrative support, etc. Hopefully, I can touch on some of the fundamental areas that need to be addressed in every setting.

INFORMATION

The information part of the puzzle can be one of the most vexing and, if done right, the most gratifying. A strong microbiology based Information System or a Laboratory Information System (LIS) is required, as well as an interactive link between the various hospital information systems and the LIS. The larger and more complex the institution, the larger and more complex the computer system that will be needed. There are several microbiology based computer systems that can be used to link sites and can provide the necessary microbiological information flow. However, the movement of the necessary hospital specific test requisitions, billing and patient report information is the challenge. It is likely that each institution will have different billing codes, some different testing requirements and different antimicrobial formularies. The presently available computer systems that provide up-to-date information to the off-site institution and generate hospital specific reports, i.e. different formats using the appropriate formulary, are not all that common. Another major consideration is the short-term and long-term storage of patient information to be used to generate institution and unit specific antibiograms, infection control and QA studies, population studies and, the need to retrieve specific patient records. Some examples of systems that are based in the microbiology laboratory are Lancraft, Inc. which supplies the software for Analytab Products, Inc. (API) and Microbiological Specialty Inc. (MSI). Two other major companies with products that should be considered would be Vitek and the Microscan System from Baxter. The companies involved in stand alone and local area network (LAN) LIS software have been reviewed in CAP Today.[5] As one can guess, some of these systems suit different needs and many do not have a strong microbiology component. The choices of systems to fulfill the need are varied. The need is inescapable.

TRANSPORTATION

Moving specimens or cultures from a satellite laboratory to a centralized or main laboratory is a seemingly simple concept. As opposed to the computerization of a laboratory, which can be painful but ultimately rewarding, this simple concept of moving materials from point A to point B is fraught with problems, chronic and acute, sporadic and continuous. Depending on the personnel that one hires for this job and one's diligence in training and providing continuing education to them, the ability to effectively move materials that may need to remain frozen, or need to be refrigerated not frozen, or require anaerobic conditions, may be greatly impaired. In addition, some specimens need to be STAT, some sort-of-STAT, and some specimens have minimal requirements. Managers would like to have one delivery per day; microbiologists would like one every hour, 24 hours per day. The compromise is somewhere in between. Once per day transportation is inadequate in most settings and every hour transportation probably requires more personnel to move the specimens than were needed originally to handle the specimens. Although most microbiology tests are not STAT tests, arrangements need to be made for those specimens, 24 hours per day, that do need STAT transportation.

"To plate or not to plate, that is the question. Whether it is nobler to . . . " - a butchered quote but, an important concept in centralizing laboratory services. The level of service that is provided by the microbiology laboratory at each of the sites can vary from STAT labs with generalists providing the minimum of microbiology, to essentially full-service laboratories at each site with only the esoteric tests sent to the main laboratory. The more centralized the services, the more efficient the laboratory can be operationally, i.e. cheaper to operate. The management usually supports the latter, the minimalists concept, and the medical staffs and the microbiologists usually favor the former. What is the solution? Again, there are no revolutionary answers. Basically, what is needed is careful planning, and someone responsible and empowered to address the somewhat complex and changing issues. Finally, as this symposium has stressed again and again, the process must be interactive.

EDUCATION/INTERACTIONS

It is not a new concept to say that the director and the supervisor of the microbiology laboratory need to know what is going on in their clinical laboratory but, the farther flung the microbiology laboratory becomes, the more important it becomes that the people responsible for the laboratory strive to keep in touch with the bench technologists on a day-to-day basis. There also needs to be a mechanism for the satellite laboratories to easily and conveniently interact with the supervisor and director on an informal and ongoing basis. Beyond these informal day-to-day interactions, there needs to be a formal, structured system to allow the satellite laboratories, the main laboratory and the medical staffs at all of the involved institutions to reach both professional and operational support. Some laboratories have instituted laboratory rounds where the supervisor and/or director review cultures and answer questions and receive comments from the bench-level technologists at a particular time every day. Designated telephone lines for the satellite laboratories and the appropriate medical staffs from those satellites have also been effective. Most directors carry pagers. I would propose that directors and supervisors must be available at reasonable times of the day, seven days per week, to answer questions, offer assistance and, in general, respond to the needs of the satellite laboratories and their medical staff.

Formal inservice programs are a requirement of JCAHO and CAP. Again, the concept is not new. As the network expands, the need for inservices expands. Not only does the need for inservices expand but the personnel that need to be involved expands. Formal inservice programs should be established for the main laboratory and the satellite laboratories. At least initially, new tests will be introduced rapidly, procedures will need to be standardized, life will change fast and everyone will need to be involved, and kept current. In addition, as test menus are modified or new methods are introduced, the people responsible for collecting the specimens must be part of the process, be they nursing or phlebotomy. Finally, the hospital committees that overlap with the microbiology laboratory at any and all of the sites involved must be a part of the process. The microbiology laboratory needs to be represented by the director, supervisor or designate on the Pharmacy and Therapeutics Committees, the Formulary Committees, and the Infection Control Committee of each hospital involved. The benefits from this involvement can be and should be bi-directional. The doctorate level or supervisory level personnel from the main laboratory should be able to contribute to these committees as well as represent the consolidated laboratory needs and overall direction. The consolidated laboratory can clearly state the concerns and needs of the medical staff and the hospital personnel for whom they provide the service.

Finally, formal meetings with the management at each site are necessary to keep everyone informed and must become part of the process. The managers of the satellite laboratories will need to know about changes and what these changes may mean to their budget. In addition, they may well be able to offer constructive input into the operation. Other formal meetings that are needed include meetings with the pathologists at the satellite laboratories. A formal interaction with the medical staff at the satellite hospitals is imperative. The director of the main microbiology laboratory should attend some of the medical staff meetings at each hospital as well present or be part of a presentation to those medical staff members whenever possible.

"Is one lab in town enough?" Maybe. The laboratory and the necessary administrations must be willing to commit the time and resources to make it work to the benefit of the institutions, medical staffs and patients. Microbiology services can be centralized, services improved and costs reduced but, the process can only be considered successful if transportation, information and education/interaction issues are examined and answered completely. If the driving force and single focus is money, these efforts will surely fail.

REFERENCES

1. S. H. Katz, B. Diamond, J. E. Prier, T. H. Lukaszczyk, and J. B. Kregerreis, Regionalization of Laboratory Services, Health Lab Science 10:787 (1973).
2. K. Gal and A. Hanok, Saving through Centralization, J. Amer. Hosp. Assoc. 44:60 (1970).
3. J. E. Prier, Regional Laboratory Services, Computers in Laboratory Medicine, A symposium Presentation, 1975.
4. B. C. Vladeck, Hospitals and the Public Purse, Transactions and Studies of the College of Physicians of Philadelphia 12:263 (1990).
5. R. D. Aller, M. Weilert, and O. G. Pasia, More LIS Installed is Not Necessarily Better, CAP Today 5:52 (1991).

THE FDA REVIEW CRITERIA FOR ASSESSMENT OF ANTIMICROBIAL SUSCEPTIBILITY DEVICES - TOO MUCH OR NOT ENOUGH REGULATION?

Karla M. Tomfohrde

Microscan Division, Baxter Diagnostics
West Sacramento, California

INTRODUCTION

The Medical Device Law of 1976 provided specific regulations for use by the Food and Drug Administration (FDA) in the evaluation of substantial equivalency and safety and effectiveness of medical devices, including *in vitro* diagnostic devices such as antimicrobial susceptibility testing systems, prior to clearance for market release. As technology progressed through the years, the FDA continued to refine the requirements used in their evaluation of new devices to insure that each new technological advance was thoroughly tested prior to market introduction. Recently, the FDA Office of Device Evaluation published a document entitled, "Review Criteria for Assessment of Antimicrobial Susceptibility Devices". [1] This document details the FDA recommendations for the type of clinical testing a manufacturer should perform; the way data should be presented to the FDA; and the minimal performance characteristics of a new system. These new review criteria will be discussed along with their potential implications for the manufacturers of Antimicrobial Susceptibility Test(AST) devices.

Types of Devices

Three types of antimicrobial susceptibility devices are currently marketed in the United States. These are:

1. Disks or strips containing antimicrobial agents for agar diffusion testing.

2. Microdilution panels containing either a range of serial, two-fold dilution of antimicrobial agents for the determination of the minimum inhibitory concentration (MIC) or selected dilutions (usually 3 or less) for determination of categorical results.

3. Automated, semi-automated and manual systems using various, non-traditional methods such as shortened incubation periods and growth-rate interpolation.

Antimicrobial Susceptibility Testing, Edited by
J.A. Poupard *et al.*, Plenum Press, New York, 1994

The FDA requires a different type of submission depending upon the general type of device. Either a premarket approval application(PMA) or a 510(k) is required. A PMA provides the FDA with performance data to verify the safety and effectiveness of the new device. A 510(k) contains performance data to show that a new device is substantially equivalent to an existing device. For antimicrobial susceptibility devices, the type of submission is based primarily on the length of incubation. Table 1 defines these submissions and lists examples of the commercial systems requiring each type of submission.

Clinical Trials

The performance characteristics of a new antimicrobial susceptibility device or the performance characteristics of a new antimicrobial agent available in a currently marketed device are determined through testing in clinical microbiology laboratories. Comparative testing of the new device/drug to an accepted reference method (generally those of the National Committee for Clinical Laboratory Standards - NCCLS) is done to determine clinical efficacy, reproducibility, efficacy with challenge strains from the Centers for Disease Control (CDC), stability, and Quality Control (QC) ranges. [2,3,4] The organisms selected for this comparative testing should include strains for which the antimicrobial agents included in the device have a spectrum of activity and should include both susceptible and resistant isolates. [5,6,7] Tables 2 and 3 detail the minimum testing the FDA recommends for new or currently marketed devices.

The results of comparative testing comprise the bulk of the data analyzed by the FDA to establish the substantial equivalency or to determine the safety and effectiveness of the new device/drug. The efficacy data with the clinical isolates and CDC challenge organisms are analyzed to determine the essential agreement (EA) for devices containing full doubling dilution MIC formats or categorical agreement (CA) for devices not utilizing full MIC formats (e.g. breakpoint formats). EA occurs when the result of the test device agrees exactly with or is within ± one dilution of the reference system result. CA occurs when the test device and reference system results agree with each other when using the NCCLS or FDA interpretive criteria. CA is also used to determine the interpretive error (see Table 4 for definition of errors) for full MIC systems with test device results which are ≥ two doubling dilutions from the reference method results. Data are presented as shown in Table 5 for each genus/species tests with each antimicrobial agent in the test device.

In order to insure that the errors seen with a new device/drug are related to the new device/drug rather than the reference method or the organism under test, repeat testing of the organism/drug combinations with major or very major errors is encouraged. Testing is performed in triplicate on both the test device and reference method using the same inocula. The actual MIC results for both systems and the final errors are summarized and the original efficacy or challenge strains efficacy results are edited.

Reproducibility and QC strain testing is performed using organisms with known MIC results. Reproducibility is determined by testing each organism in triplicate on the test device and the reference system for three consecutive days at each of the clinical trial sites. QC testing is performed on each day of the clinical trial. Both the test device and reference system are to be tested.

Table 1
Types of FDA Submissions Required for Antimicrobial Susceptibility Testing Devices

Type of Submission	Device Characteristics	Commercial Systems
Premarket Approval (PMA)	Any device not based on traditional overnight incubation (< 16 hours)	Baxter MicroScan - Rapid Fluorogenic Panel Organon Teknika Corp - Autobac Series II Vitek Systems - AutoMicrobic System
510(k) Premarket Notification with Performance Data	(1) Microdilution MIC or MIC breakpoint systems using overnight (16-24 hours) incubation	Analytab Products - UniScept Baxter MicroScan Becton Dickinson Co. - Sceptor Radiometer of America - Sensititre
	(2) Non-traditional format systems using overnight incubation but employing dilutions schemes other than broth or agar dilution (e.g. density gradients)	AB Biodisk - E Test
510 (k) Premarket notification without performance data	Antimicrobial disks	Becton Dickinson Co. Difco

Table 2
***Original Submissions* for Antimicrobial Susceptibility Devices**

		<16 hour Incubation System	16 - 24 hour Incubation	Kirby-Bauer disks
FDA Required Submission:		**PMA**	**510(K)**	**510(K)**
ITEMS (Minimum)		All	Serial dilutions (MIC) [e]	All concentrations
Number of Sites		3	3	1
Organisms	Fresh Clinical[a]	300/site	100/site	0
	CDC Challenge[b]	100/site	75/site	0
	Stock[c]	100/site	50/site	0
Reproducibility Matrix		10X3X3X3[d]	10X3X3X3[d]	0
Interpretive Breakpoint		FDA/NCCLS	FDA/NCCLS	FDA
Stability (3 lots)		real time	real time	real time
QC Reference and Test Panel Results	NCCLS Strains	yes	yes	yes
	(Other Mfg. Recommended)	optional	optional	NA
	On-scale	at least 1	at least 1	NCCLS organisms
NCCLS Reference Method		MIC or KB	MIC or KB	KB

Note: see Table 3 for footnotes.

Source: Review Criteria for Assessment of Antimicrobial Susceptibility Devices, May 1990.

Table 3

Additional Submissions **to Marketed Antimicrobial Susceptibility Device**

Items[f]		< 16 hour Incubation Systems	16-24 hour Incubation Systems
			Serial Dilutions (MIC)[e]
Number of Sites (including in-house as one site)		3	2
Organisms	Fresh Clinical[a]	100/site	100/site
	CDC Challenge[b]	75/site	50/site
	Stock[c]	as appropriate	as appropriate
Reproducibility Matrix		10X3X3X3[d]	10X3X3X2[d]

a) Organisms tested should be representative of the antimicrobial's specturm of activity.

1) When testing expanded spectrum penicillins and cephalosporins, include at least 10 resistant isolates each of the following species: *Enterobacter spp., Citrobacter freundi, Serratia Marcesans, Pseudomonas aeruginosa.*

2) When testing other types of antimicrobial, include at leaset 20 organisms representative of each known mode of resistance for that antimicrobial.

3) When testing *Staphylococcus spp.*, include 25 oxacillin susceptible and 25 oxacillin resistant strains of both coagulase positive coagulase negative strains (100 total strains).

b) The same CDC challenge strains are to be tested at all sites.

c) Stock organisms can be selected from each site to supplement numbers of infrequently isolated species or to represent endemic resistant strains.

d) Include at least 5 on-scale organisms; QC organisms must be included. 10 strains, 3 sites, 3 days, in triplicate.

e) Breakpoint selection should be chosen and extracted from full MIC dilution data.

f) All other items are the same as for original submissions (refer to Table 2).

Source: Review Criteria for Assessment of Antimicrobial Susceptibility Devices, May 1990.

Minimal Performance Characteristics

The clinical utility of a susceptibility test result is dependent on the accuracy of the susceptibility system generating the result as well as the ability of the system to detect both susceptibility and resistance in organisms of clinical significance. The accuracy is measured by the EA or CA. The ability to detect susceptibility is assessed by the major error rate while the ability to detect resistance is assessed by the very major error rate. The FDA has established performance characteristics for EA/CA and major/very major errors as detailed below:

1. EA or CA should be > 90% for each organism/drug

2. Major errors should be < 3% for the susceptible strains

3. Very major errors should be > 1.5% for the resistant strains with similar mechanisms of resistance [6,8]

If any of these criteria are not met, an alternate method for testing must be recommended in the product labeling.

The ability of a new device to detect resistance is dependent on the presence of resistant strains in the general environment. A minimum of 20 strains representing each mechanism of resistance for each antimicrobial agent must be tested. If sufficient numbers of resistant isolates are not tested, the "warning" statement given below must be included in the product labeling.

"The ability of the ABC system to detect resistance to ("Antimicrobic") among the Enterobacteriaceae (or other organisms) is unknown because resistant strains were not available at the time of comparative testing." [1]

If and when resistance develops, additional testing and another submission to the FDA will be required before this statement can be removed from the labeling.

Additional performance characteristics include the determination of growth failure rates and comparison of QC ranges to those of the NCCLS. Growth failure rates exceeding 10% for any genus or species and any variance from the NCCLS expected QC ranges must be described in the labeling.

Implications for Manufacturers

The manufacturers of the antimicrobial susceptibility devices concur with the purpose of the FDA Review Criteria document, namely:

> "To ensure well-standardized, reliable, and reproducible commercially available tests for determining the *in vitro* susceptibility of infectious bacteria" [1]

However, some of the specific performance criteria raise questions which must still be answered and which could negatively impact future development by AST manufacturers.

Table 4
Definition of Interpretive Categorial Errors

Error	Test Device Result	Reference Result
Minor	a. Resistant (R) or Susceptible (S)	a. Intermediate (I) or Moderately Susceptible (MS)
	b. I or MS	b. R or S
Major	R	S
Very Major	S	R

First, both major and very major errors have always been considered significant and the percentage of these errors has been critically evaluated. In 1982, Sherris and Ryan proposed guidelines for acceptable accuracy of bacterial identification and antibiotic test procedures which have been used by manufacturers in assessing new devices. Their recommendations included:

1. The percentage of very major errors attributable to the new procedure should be less than 1.5% for all individual species to be tested.

2. The overall percentages of errors attributable to the new procedure should not exceed 5% in tests on random clinical isolates.[9]

Gradus et al. revised these recommendations in 1985 to ≤1% very major errors and ≤4% major errors for each organism/drug combination. [10]

Both of these recommendations determine the error rate based on all strains of a species tested, not the resistant or susceptible strains only. This difference is significant. For example, if 100 *E. coli* were tested (80 - S and 20 - R) and three major errors and one very major error occurred, calculating errors using all strain results in 3% major and 1% very major errors while these errors shift to 3.75% and 5% respectively when only S or R strains are used in the calculation. The very major error rate has increased five-fold. And, using the new FDA criteria, limitations would be required in the product labeling. Is this significant? It certainly will create questions in the minds of the clinical microbiologists which the manufacturers will have to answer.

Second, the majority of manufacturers are now concentrating on the development of new antimicrobial agents for their devices. These agents have little or no resistance to date for the organisms with clinical applicability. So, how will manufacturers prove their ability to detect resistance? At initial release of a new drug, a "warning" statement will probably be included in the product labeling. As resistance occurs, new clinical trials will have to be performed and a new submission made to the FDA in order to remove this "warning". This will take time and will cost money!

Third, the overall effect of these new FDA review criteria may be the delayed market introduction of new devices or the delayed addition of new drugs onto currently marketed devices as a result of more extensive clinical trial testing.

Table 5
Example of Reporting Format for Efficacy and CDC Challenge Organisms Data

Interpretive Criteria (mcg/ml):	S	I, MS	R	Antimicrobial: ______________
Reference:	____	______	____	
Test:	____	______	____	

	Reference Panel		Test Panel Results: EA or CA [b]		Interpretive Category Discrepancies [c]					
					Minor [d]		Major [e]		VeryMajor [f]	
Organisms [a]	Results	#Tested	#	%	#	%	#	%	#	%
Enterobacter sp.	Total	102	96	94	4	3.9				
	S	16	15				0	0		0
	MS/I	4	1							
	R	82	80						1	1.2
E. coli	Total	50	48	96	1	2.0				
	S	48	47				1	2.1		
	MS/I	2	1							
	R	0							0	0
TOTALS:	TOTAL	152	144	95	5	3.3	1	1.6	1	1.2
	S	64								
	MS/I	6								
	R	82								

a Organisms tested for which drug has clinical utility should be highlighted.

b EA with MIC format drugs; CA with non-MIC format drugs.

c For all test results $\geq \pm$ 2 dilutions from reference result, apply NCCLS/FDA interpretive criteria to determine Minor, Major, and Very Major errors:

d % Minor Errors = $\frac{\text{\# Minor Errors}}{\text{total strains tested}}$ X 100

e % Major Errors = $\frac{\text{\# Major Errors}}{\text{total } \underline{S} \text{ strains}}$ X 100

f % Very Major Errors = $\frac{\text{\# Very Major Errors}}{\text{total } \underline{R} \text{ strains}}$ X 100

Source: Review Criteria for Assessment of Antimicrobial Susceptibility Devices, May 1990.

For example, prior to the new criteria, a typical clinical trial consisted of 200 strains for efficacy, 50 CDC challenge strains and 25 reproducibility strains at each site. This testing would take approximately four weeks. Utilizing the new criteria with a concerted effort to include the required number of resistant strains, a clinical trial could take 10 to 12 weeks and include up to 500 efficacy strains, 100 CDC challenge organisms, 10 strains for reproducibility and daily testing of the QC organisms. In this example, the time for a clinical trial has almost tripled. The 510(k) or PMA submission would also be delayed by this same time period.

Future Considerations

In the future, close cooperation between the AST manufacturers and pharmaceutical companies will be critical to ensure that development of new drugs for AST systems is done in a timely fashion. The early establishment of QC ranges by the pharmaceutical companies and the approval of these ranges by the NCCLS subcommittee on Antimicrobial Susceptibility Testing is an essential first step. This will allow the AST manufacturers to begin their development while the pharmaceutical company's new drug application (NDA) is being reviewed by the FDA.

Additionally, while the pharmaceutical companies are gathering clinical data for the NDA, all strains which show resistance should be collected for future use by the AST manufacturers. This should ensure that those resistant strains which do exist are available for future testing.

Finally, the AST manufacturers (as well as all other interested parties) must maintain an open dialogue with the FDA. Only through continued discussions about the Review Criteria document will it evolve to a document that is truly useful to all - FDA, AST manufacturers, and clinical microbiology laboratories.

REFERENCES

1. Review Criteria for Assessment of Antimicrobial Susceptibility Devices, Food and Drug Administration (May, 1991).
2. National Committee for clinical Laboratory Standards Approved Standard M2-A4, Performance Standard for Antimicrobial Disk Susceptibility Tests, Fourth Edition, NCCLS, Villanova, PA (1990).
3. National Committee for Clinical Laboratory Standards Approved Standard M7-A2, Methods for Dilution Antimicrobial Susceptibility Tests for Bacteria that Grow Aerobically, Second Edition, NCCLS, Villanova, PA (1990).
4. National Committee for Clinical Laboratory Standards, Tentative Guideline, M23-T, Development of in vitro Susceptibility Testing Criteria and Quality Control Parameters, NCCLS, Villanova, PA (1989).
5. G.L. Mandel, R.G. Douglas, and J.E. Bennett, Principals and Practices of Infectious Diseases, 3rd Edition, Churchill Livingston, New York, pp. 218-349 (1990).
6. J.O. Yao, R.C. Moellering, Antimicrobial Agents, in: "Manual of Clinical Microbiology", 5th Edition, ASM, Washington, D.C. 1065-1098 (1991).
7. Physicians Desk Reference, 45th Edition, Medical Economics Company, Oradell, New Jersey (1991).
8. G.A. Jacoby, and G.L. Archer, New Mechanisms of Bacterial Resistance to Antimicrobial Agents, N. Engl. J. Med. 324: 601-609 (1991).
9. J.C. Sherris, and K.J. Ryan, Evaluation of Automated and Rapid Methods, in "Rapid Methods and Automation in Microbiology", R.C. Tilton, ed., ASM, Washington, D.C. (1982).
10. M.S. Gradus, Antimicrobial Susceptibility Testing Systems Past and Present, A documented overview, Part II, The Antimicrobial Newsletter, 2:73-82 (1985).

THE EVOLUTION OF CLINICAL LABORATORY REGULATION - A PRIMER FOR UNIVERSAL HEALTH CARE

James E. Prier

Philadelphia College of Osteopathic Medicine
Philadelphia, Pennsylvania

INTRODUCTION

For the past 300 years science has been providing the discovery necessary to develop a complex system of medical diagnosis, prevention and treatment. However, in the process of applying such a system to a societal need in this nation, something has been lost. This is the concept that freedom implies equal availability for all to the products and services of the medical marketplace in the sense of a "positive right", rather than the negative protection of the Bill of Rights.

It is not a matter of election that regulation of health services is fast gaining momentum, and comprehensive health care is becoming the darling of political schemes. Rather, it is an enforced reality, born of too many years of both political neglect, and irresponsibility and greed of private providers of medical services. The revelation that health care is unattainable for many may be startlingly new in the Halls of Congress, but not to those for whom the quiet shadows of each night are disturbed by the apocalyptic riders who challenge survival.

And so, the governors in conference last August loudly deplored the crisis in medical care and demanded reform, legislators at all levels have plans for remedy, and medical and public health organization have contrived multiple cures. And all have a distinctive sterility for they avoid the obvious, which is that a business in bankruptcy must turn to a single action first: that is to reduce operating costs. In the case of health services, this must be physician and hospital costs and insurance fees, and there is no alternative to the primary of this option.

Voluntary resolution of the health care crisis is not feasible. Clinical laboratory regulation is a prototype which demonstrates the conclusion that a federalist approach to health care is mandatory. It is a centrally controlled service that has evolved so as to consolidate its diverse components and finally to achieve a degree of uniformity, the result

Antimicrobial Susceptibility Testing, Edited by
J.A. Poupard *et al.*, Plenum Press, New York, 1994

being a standardized "product" at a minimum consumer cost with an optimum of quality, and, therefore, unique among the health services. The development of laboratory regulation has taken nearly 50 years and has been in experimentation most of this time. The final results of this great experiment are yet to be determined, particularly as expanded to the larger parameters of total health care. Many years ago we suggested "the optimum in rule and regulation remains as elusive as the mistress of truth, and as vague as a theory undefined."

Origin of the Clinical Laboratory

Laboratory medicine originated as an offspring to anatomic pathology in the last decades of the 19th century.[1] During that period, pathology was an integral part of clinical medicine and clinician-pathologists such as Osler, Pepper, Cushing and Welch were instrumental in moving the microscope from the medical school to the hospital. In Boston, the Massachusetts General Hospital developed microscopy and chemistry as part of the routine activities, and in Philadelphia, Pepper honored his father by establishing the William Pepper Laboratory of Clinical Medicine at the University of Pennsylvania Hospital. In Baltimore, the newly formed Johns Hopkins Hospital included a clinical laboratory, primarily for microscopic study of blood cells and malarial parasites.

Subsequent growth of laboratory medicine was first the result of developments in pathogenic bacteriology. Studies in research laboratories yielded the diagnostic tools for defining infectious diseases. The clinical application of microbiologic procedures increased rapidly over the first half of the 20th century. Immunologic techniques were later additions to the laboratory testing armamentarium, first with a variety of tests for syphilis. Clinical chemistry was a relative late-comer. Studies on urine relating to specific disease were the early application, with blood chemistry developing as routine testing during the second and third decades of this century.

The application of testing at first was a retrospective study, and used as an adjunct to necropsy examinations. Causes of death could be defined more accurately and gross and histologic changes described in terms of pathogenicity. Later the dynamic aspects of physiologic alterations in the living subject were measured by examination of tissue and fluids.

A corollary development was the reform of medical education during the first two decades of the 20th century.[2] With changes in academic programs, there was an emphasis on research and laboratory studies, and greater application of such disciplines to clinical evaluation. In addition, the advantages of associating medical schools with universities established a scientific base to medical education and postdoctoral training.

The Development of Regulation in Laboratory Medicine

As the use of laboratory analysis became an essential part of clinical evaluation, there was a proliferation of laboratory facilities, both within and outside of hospitals. And not least in importance was that they were significant profit centers within the medical enterprise. Even physicians discovered that they could make more money by conducting simple tests in their offices. Test protocols were followed but quality and accuracy were not mandatory parts of the testing process.

In 1947 Belk and Sunderman[3] reported that all was not well in the clinical laboratory. Comparative examinations of clinical chemistry analyses suggested that clinical guesswork might be a more efficient method of evaluating the state of patient health. Investigations by others, including the College of American Pathologists substantiated these doubts. In 1951 the Commonwealth of Pennsylvania enacted the Analytical-Biochemical-Biological Laboratory Act which regulated laboratories that were independent of hospitals and physicians' offices. The State of New York previously had developed a system of local laboratories with supervision over certain testing procedures, particularly serology and microbiology, and in 1964 enacted the Laboratory Services Act requiring licensure.

Studies by the New York City Public Health Department, under the leadership of Morris Schaeffer in the 1960's, demonstrated a significant level of error in phases of clinical laboratory testing.[4] This created national publicity which came to the attention of State and Federal Agencies and Congress. in 1961, the Pennsylvania legislature responded to the reports of testing inaccuracies by requiring proficiency testing of all laboratories under the State's jurisdiction. In the private sector, significant self-regulating programs were started, with the subsequent result of highly efficient proficiency testing systems such as those now conducted by the College of American Pathologists and the American Association of Bioanalysts. States and New York City also developed testing for regulatory purposes.

Until the 1960s, clinical laboratory regulation was confined mostly to the states in accord with their delegated police powers. Federal regulations for health related matters were minimal, confined to specific limited statutory provisions. The Social Security Act of 1935 provided medical assistance to the blind and disabled; amendments in 1950 provided direct payments to vendors of medical services; in 1960 the Kerr-Mills Bill established medical assistance to the aged, leading to the 1965 amendments to the Social Security Act which created Medicare and aid to the medically indigent. Clinical laboratories which provided services for recipients of Medicare and Medicaid were included, with hospitals under comprehensive provisions and independent laboratories under a separate regulatory format. Physicians' office laboratories continued to be exempt. The next federal action was the amendment to the Public Health Service Act with a statute called the Clinical Laboratory Improvement Act of 1967. The laboratories affected were those that performed analyses on a minimal number of specimens transported across the state borders.

Since states were the agents for administration of the Medicare regulations, each became involved in matters of clinical laboratory evaluation. Subsequently a number of states developed their own statutes and regulations, some of which are equal to or more stringent than the federal medicare rules.

In 1972, a significant amendment to the Pennsylvania Clinical Laboratory Act was enacted by the General Assembly.[5] All exemptions except federal laboratories were removed. The State Health Department declared by regulation that physician office laboratories were subject to the requirements of the Act. In response, the Pennsylvania Medical Society brought a suit in equity to enjoin the enforcement of regulations and to remove physicians from the jurisdiction of the Act. The case reached the State Supreme Court, and in 1977 it was ruled that laboratory work in physicians' offices was not outside the intent of the legislation. The conclusion to be drawn from this case as well as a previous New York case (1966), and the 1969 CAP consent decree, clearly indicated the legal concept that the practice of clinical medicine and the practice of laboratory medicine are separate and distinct, requiring different talents and educational and experience qualifications.

In 1988, as a result of negative press coverage regarding examinations for cervical carcinoma and other laboratory tests, Congress enacted an amendment to CLIA '67 which is referred to as the Clinical Laboratory Improvement Amendments of 1988 (CLIA'88). The structure and intent of the statute clearly follow the prototype of the Pennsylvania Clinical Laboratory Act. Subsequent proposed regulations are consistent with this comparison. For example, the Pennsylvania regulations established 3 levels of testing as did the CLIA '88 proposals. In both cases, the ability of laboratory work to be done by a physician on material from his own patient is severely restricted.

> "...but the law has compulsive power, while it is at the same time a rule proceeding from a sort of practical wisdom and reason."
>
> Aristotle (on ethics)

The Elements of CLIA '88

It would be redundant for this audience to detail provisions of CLIA '88 and its proposed regulations. They have been discussed in every relevant journal and are the cause of continuous debate.

The language of the statute clearly indicates that all laboratories, regardless of site and ownership, are to be covered equally. The proposed regulations, in part at least, are consistent with this intent. Specific areas of coverage are personnel, performance standards, inspections, proficiency testing, fees and sanctions. Proposed regulations have been published in several sets and there is no doubt that these have captured the attention of the laboratory community. Approximately 60,000 responses were received by the Health Care Financing Administration regarding one or more features of the proposals. Many of these, of course, are the products of organized letter writing campaigns fostered by private interest groups. The writing of CLIA '88 regulations has been assigned to the Health Care Financing Administration (HCFA) and the Centers for Disease Control (CDC), with consultation in some areas from the Food and Drug Administration (FDA). Advance reports regarding the final regulations have confirmed the fears of many, that the pressures of corporate interests with potential markets, particularly in physician office laboratories, have caused the regulations to be inconsistent in part with the statutory intent. The result may be that quality will be diminished, unnecessary testing will increase, and costs will not be contained. This will be particularly true if physician office laboratories are permitted to continue and expand with minimal personnel requirements, and inadequate quality control. It should also be noted that the three federal agencies assigned the task of constructing these specific regulations are devoid of competency to achieve this objective.

Clinical Laboratory Regulation in Perspective

It may be well for a moment to view in the mirror of the past the matter of laboratory regulations. As it should be, each step along the way has been a response to a demonstrated need for improving a critical public health service. It seems to have been an orderly progression, but as with all systems of government that evolve to meet changing societal needs and demands, so too must the delivery of health services be constrained by regulation only to the degree consistent with optimal public benefit, and not to an extent that is unnecessarily oppressive.

The history of clinical laboratory regulation is a microcosm of sorts; a prototype which might be applicable to other aspects of health services. There seems to be general agreement that medical care in the U.S. is becoming an unaffordable commodity. There was a time when organized medicine as a private enterprise could have disciplined itself so that its services might be provided at reasonable cost, but this time has gone. The people of the nation and the congress that represents them have said, "A plague upon your house; we shall take from you the public responsibility which is inherent in your licenses to operate; and with that, the freedoms enjoyed."

Members of congress are examining the systems of other countries such as Canada, Britain, and Germany. The AMA is admitting that a problem exists, and seeks a remedy that does not include constraints of fees, and state agencies look first for political acceptability. The lesson from the laboratory experience is reliable: direct central control without a complex use of third parties is the effective option providing that cost caps are strictly enforced, particularly of physician and hospital charges.

Change in the traditional method of health care delivery will not be easy. As in the case of laboratory regulation, the adversely affected providers already are girding for battle. Their weapons may not be adequate though, for they are already chipped and broken from defeats upon the battlefields of Medicare, malpractice, and investment regulation.

The Future of Clinical Laboratories

As with any law pertaining to a dynamic subject, CLIA '88 is neither the ultimate nor the end of clinical laboratory regulation development. There will be additional changes in laboratory regulation in the future, as there will for all other segments of the health delivery system. The significant difference is that the regulatory mechanism is in place for the laboratories, but the remainder of health services must yet catch up to the model.

It can be expected that the fee schedules that have been developed for laboratory work done for federal programs will be extended to cover all private recipients of laboratory services. Also, review and audit of tests performed will be applied in order to limit the amount of unnecessary testing being done, both in hospitals and for outpatients. Particular consideration will be given to procedures that have high unit costs, but inconsistent clinical application, such as antibiotic susceptibility testing.

Costs of some tests may be "rolled-in" as part of a general examination fee. Also, within the next few years there will be a marked decline in physician office laboratories, and possibly the elimination of many of these sites by state regulation. Although a limited profit center is now associated with POLs, future restrictions may eliminate this reason for existence. Physician ownership of independent laboratories already is being eliminated by regulation, as is ownership in other medical facilities. This area will extend from federally supported patients to all recipients.

It is probable that a significant increase in state regulatory programs will occur in the near future. State agencies can now be designated as contractors for implementation of the federal programs, but states will seek to obtain preemption through more stringent rules. They will see the potential of fiscal transfer from federal budget as an advantage to their administrative status.

Proficiency testing of licensed laboratories is mandated by statute. It can be anticipated that data collected, as now required, if done by appropriate method, will result in changes. My opinion is that it will be shown that current requirements are excessive and that much less can be done at less cost for adequate quality assurance. The optimal amount of proficiency testing never has been determined, and this issue was placed into question 20 years ago. [6,7]

Although the mega-laboratories may continue to expand for a short period, there will be a resurgence of small community laboratories. There are several reasons. One is that a decline in POLs will not be associated with a financial benefit to physician referral, so many will prefer local service, particularly when the patient can be directed to the laboratory. Decreasing volume and in some cases decreasing complexity may modify the need for expensive high volume machinery, and cost-effectiveness may shift work into smaller and less expensive operations.

There will be increasing control by the patient over his laboratory testing. The requirement that only prescription testing can be done will no longer be in effect, and the patient will assume the right to initiate test requests for himself, and he will become more astute in such matters. He may even negate some testing that is requested by his physician. The laboratory will not have the option to refuse patient requests, even in hospitals.

Personnel changes can be expected, particularly in areas of supervision. Less emphasis on academic qualifications and more on technical training and experience will be determinative. In the cases of both technologists and supervisors a specific degree will not be either a barrier to or a criterion for advancement. For director positions probably the present moderately complex proposals and the criteria of the Joint Commission for the Accreditation of Healthcare Organizations will control.

Although automated chemistry systems are economical in routine laboratory analyses, the same may not be true of microbiology systems unless very large volumes ares involved, and perhaps not even then. It is predicted that automation in microbiology, because of more critical test selection may decrease significantly. Screening will be more selective and cost a significant factor in antibiotic selection. But, eventually, cost control of products of the pharmaceutical industry, an ultimate necessity, may reverse this trend.

The administration of the CLIA'88 is still in question, in spite of the quality and comprehensiveness of the statute and regulatory proposals. This is part of the agonizing process of moving the legislative intent into administrative function, a factor also prototypic of all past and future health service by government.

So, after the arguments have been aired, the debate concluded, the damaged egos assuaged and the regulations finalized, there remains this dilemma: how will a program so vast in its concepts be administered? The greatest fear of all is that the federal government will attempt this task directly, and thus doom a positive step in improving an essential medical service to ignominious obscurity. The statute does permit the establishment of agencies to administer the regulatory process. However, there can be little doubt that there is but a single class of such agency that is sufficiently capable of accepting the task. This is the state administrators who have been supervising the Medicare and other laboratory programs for the past 20 years. If there should be a decision not to delegate solely to those cadre of state officials, all in federal laboratory regulation may become an academic exercise, devoid of substance. The concept that private agencies can provide such services, particularly those with vested commercial interest, is merely ludicrous.

In the administrative ranks, as in the laboratory itself, success relates to competency and adequate numbers of persons. In each there now are deficiencies. Unless this is corrected soon, the prize may be lost for want of both captain and crew.

When we were first plunged into the strange world of clinical laboratory regulation, there was no mentor from whom to learn. But the lessons of the past three decades have taught much to those in state government and private organizations, as well as laboratory personnel, about the complexities of sound and proper control of health services, and the proficient conduct of laboratory analyses. And the voices of those who achieved the skills of experience can still be heard from the state agencies, the laboratories, and the professional organizations. However, time moves too swiftly for us all, and my colleagues are wont to say that our time in the forum has past, and it is the youth we envy who must now accept the challenge to guard the progress of this great adventure; and to rail against government when it fails in its responsibility or oversteps its proper bounds. But to those who have grown tired, and as we all do at times, long for the quiet place, I suggest that care be taken so that the experience of a generation not be lost, and so hearken to the words of an old poet:

> "The woods are lovely, dark and deep
> But I have promises to keep,
> And miles to go before I sleep,
> And miles to go before I sleep."
>
> Robert Frost

In the new regulated environment, adequate personnel for all tasks may become the most critical issue. For a time the wisdom and experience of the old warriors may be necessary to assure success of this unprecedented step into the control of an essential health service.

> "...for I regard them as travellers who have gone a journey which, I too, may have to go, and of whom I ought to enquire whether the way is smooth and easy, or rugged and difficult."
>
> Plato

CONCLUSION

The federal clinical laboratory statute and regulations, if placed into effective use, shall keynote a new era in health service. It will be the first example of a national comprehensive system for medical care. Yet the arguments over the latter will be many, ranging from principle of mechanics. Ethical issues already have been raised, stimulated by the Oregon Basic Health Services Act of 1989 and other ethics discussions. [8,9,10] But the overriding issue is cost and unless this factor is primary, all others become academic.

It now becomes the challenge to Congress to set aside the influences and political pressures from those with interests in the business of medicine, and create a centrally controlled affordable health plan. So far this public responsibility of legislators seems elusive.

Change is not easy, and is particularly, difficult for those who have enjoyed the liberty to determine the price of the medical marketplace. For them the winds of change blow cold, and they shudder in fear of a darkening night. But for the architect of an evolving health delivery structure the issue is clear. Either the impending disaster in health care will be thwarted; or the mother of the nation will find cause to weep for her children; the old whom we revere will no longer have the option to live a bit longer, and the heirs of our time shall have a basic right denied.

> "Man must come to terms with himself. If he does not learn self-discipline, if he fails to solve the problems he has created, then at least those problems will be solved for him. The solution will lie in the hands of one or all of his age-old enemies; Famine, Pestilence and War, the three horsemen of the Apocalypse, who bring in their train the Fourth Rider, Death upon his pale horse."
>
> From: *Disease and History*
> by Cartwright

REFERENCES

1. L. S. King, Clinical Laboratories Become Important, 1870-1900. JAMA 249:3025 (1983).
2. L. S. King, The Flexner Report of 1910, JAMA 251:1079 (1984).
3. W. P. Belk, and F. W. Sunderman, A Survey of the Accuracy of Chemical Analyses in Clinical Laboratories, AMJ.CI.Path., 17:853 (1947)
4. M. Schaeffer, D. Widelock, S. Blatt and M. E. Wilson, The Clinical Laboratory Improvement Program in New York City, Health Lab.Sci., 4:72 (1967).
5. The Clinical Laboratory Act, 35 P.S., 2151-2165
6. J. E. Prier, L. Sideman, and I. J. Yankevitch, Clinical Laboratory Proficiency Testing, Health Lab. Sci., 5:12 (1968).
7. R. Bugg, Nuts and Bolts Session Probes Proficiency Testing, Lab. World, p. 1356, (Nov. 1971).
8. R. M. Veath, Should Basic Care Get Priority? Doubts About Rationing the Oregon Way, Kennedy Institute of Ethics Journal 1:187 (1991).
9. R. M. Veath, Physicians and Cost Containment, Jurimetrics, 30:461 (1990).
10. P. B. Ginsburg, Alternative Approaches to Health Care Cost Containment, Jurimetrics, 30:447 (1990).

CURRENT ISSUES IN ANTIMICROBIAL SUSCEPTIBILITY TESTING

James A. Poupard[1] and Lori R. Walsh[2]

[1]SmithKline Beecham
King of Prussia, Pennsylvania

[2]Abington Memorial Hospital
Abington, Pennsylvania

INTRODUCTION

This section is based on a point-counterpoint session designed to help in defining the relevant issues facing the microbiologist today, with particular reference to the automated test systems. These issues are a compilation of a large number of issues that were raised during the symposium and the point-counterpoint session. In the point-counterpoint session, certain issues were raised to encourage discussions of different views on both sides of the point under discussion. Seminar attendees were encouraged to respond and to express their feeling on the subjects in writing. The authors have attempted to condense these discussions and written comments into the six general subjects that follow. It should be noted that there was considerable editing to condense the material into this form.

Rapid vs. Overnight Results

One of the advantages of using some of the automated systems is the ability to generate a final result on the same day that an organism is isolated. This is an issue that has received much attention since the first automated system became available in 1974. The advantage of same day reporting is perceived as one of the reasons to invest in the more expensive systems, and is significant in the marketing of these instruments. When a microbiologist is cost justifying the outlay of a considerable sum of money for a rapid system, this issue is often at the center of the justification to a hospital or laboratory administrator. The concept of decreased turn around time leading to more efficient therapy and earlier discharge of the patient is widely quoted. This theoretically will result in a cost saving to the medical center operating in an atmosphere of DRGs and cost reductions per patient.

In the 1990's this concept is still presented as a sound advantage for automation, but the argument is not as strong as it was in the 1980's. The arguments against the need for rapid systems are receiving attention. The following are some common arguments against the need to invest considerable funds in a system that yields rapid turn around time:

1. In severally ill patients, therapeutic decisions are made on admission, and physicians will often continue treating the patient with empiric therapy for 72 hours in an effort to stabilize the patient. The differences in turn-around-time between the rapid vs. the conventional systems is often simply the difference in obtaining the results late in the day or early evening of the second day from admission rather than having the results on the morning of the third day after admission. The question is raised as to whether the cost of an automated system is justified under these circumstances.

2. Rapid often means six hours from the time of inoculation, and therefore, the results are often available in the late afternoon or early evening. Therefore, results are often generated at a time when most microbiology laboratories are under staffed or staffed by generalists.

3. There is a general concern that the more rapid the system, the more inconsistencies between certain organism/drug combinations will be found. As rapid systems become improved, this factor will probably be reduced, however, at the present time, this concept will probably persist.

4. The generation of same day results places pressure for increased staffing in the morning to enable organisms to be isolated and prepared for entry into the automated system. This is combined with the need for qualified technologists to be present at the end of the day to review results prior to release to the patient records. This situation is particularly significant in staffing microbiology laboratories on weekends.

It should be noted that in larger laboratories, these factors may be minimized, but in smaller community hospitals, this issue may be quite significant. There is a definite need for more careful studies to demonstrate the utility of rapid results in a wind variety of clinical settings. Until these studies are performed, published, and more thoroughly debated, the questions relating to this issue will remain.

Automated vs. Non-Automated Methodologies

In the past, automated systems were justified as necessary to increase efficiency. This is a key issue for the '90's. This issue has become more complex than in the past. Most microbiologists agree that automated test systems will, in all probability, not be a factor in reducing laboratory staff. However, this does not eliminate the efficiency argument, it simply redirects it. Do automated antimicrobial susceptibility test systems increase efficiency? The answer is probably "yes", but requires a full investigation of many issues. Some of these issues will be addressed under the specific sections that follow. The following are some factors relating to how automation increases efficiency. Automated systems permit:

1. Better use of data processing when the susceptibility testing instrument is placed on - line with the laboratory data processing.

2. Elimination of technologists' variability in reading results.

3. An efficient method for generating antibiograms and other epidemiology reports necessary for infection control decisions as well as formulary decisions.

4. Certain programs and algorithms that can take into account the ever increasing exceptions and rules (such as NCCLS table footnotes) for breakpoint interpretation.

5. The use of cascading programs that permit reporting of certain antimicrobial agents, such as higher generations cephalosporins, only when the first generation drugs in that family are resistant.

6. The use of software programs that can incorporate test results, such as β-lactamase production, into decision making programs based on these results. This would include altering ampicillin susceptibility results in organisms found to be β-lactamase positive, or using a series of drug results to detect inconsistencies with certain other drug results on the same test panel.

Susceptible, Intermediate or Resistant (S, I, or R) Results vs. MIC

Interest in automated susceptibility test systems was initially stimulated by a perceived desire for an MIC result instead of the familiar S, I, or R result generated by the disk diffusion method. Although this is still a significant factor, the impact of MICs has been diminished somewhat by several years of experience in reporting out these results. Many laboratories with access to MIC results find it "cumbersome" to incorporate MICs into the final laboratory report. The perception that converting the MIC result to S, I, or R is more accurate than relying on zones of inhibition for the purpose may be relevant. Some newer methods, such as the Biomic system converts zone sizes to an MIC result. Some participants raise the question that since most microbiologists are converting MICs to S, I, and R, why is an expensive system necessary to generate the same result as that obtained with disk diffusion methods? This is an issue that will probably continue to receive attention in the future. Some of these issues are addressed in the next section - MIC vs. Breakpoint Result.

MIC vs. Breakpoint Results

Some panel manufacturers offer MIC panels and Breakpoint panels. In general, to generate an MIC, multiple wells with different drug concentrations are required. This limits the number of drugs per panel. Breakpoint panels, which provide S, I, or R results, usually contain less wells per drug, and therefore, will contain more drugs per test panel. It is obvious that the more drugs a single panel contains, the more usage that panel will receive and it will better serve the needs of the microbiology department. The advantage of breakpoint panels are sometimes diminished by the lack of sensitivity in differentiating minimal differences with QC organisms, especially with those drugs that do not have an intermediate breakpoint or those drugs containing very tight-ranged intermediate breakpoints. As more experience is gained with breakpoint panels, these issues will become better defined. At this time, there is significant interest in breakpoint panels, and it is probably safe to conclude that this interest will continue. The use of these panels permit some laboratories to test all routine gram-negative isolates using a single panel. As some of these issues are addressed, more light will be shed on the MIC vs. S, I, and R debate.

Frozen vs. Dehydrated Panels

Some systems still use frozen panels. For most drugs, there are no differences in results between frozen and dehydrated panels. However, if one is considering the use of these panels, the fact that some drugs are not stable at -20° C must be considered. This creates significant limitations in the use of these panels. Preparation and shipment at -70° C solves most of these problems. However, two points should be considered relating to this issue: (i) few laboratories have sufficient -70° C freezer space, and (ii) it is difficult to produce and ship panels at -70° C. It is probably safe to predict that the use of frozen panels will decrease in the future.

Automated System Limitations

An often cited problem with the automated systems is the need to test certain organism/drug combinations by an alternate method. This is often used as an argument against switching to an automated system. Although this argument is valid, as noted previously, some newer systems such as the E-test, offer reasonable alternatives for testing isolates external to the main test system. Organisms such as *Haemophilus influenzae* and more other fastidious organisms are often cited as creating significant problems for the automated systems. Two points should be made relating to this subject: (i) the availability of methods to be used as supplements to the automated systems will probably increase in the future, and (ii) instrument manufacturers are making a concerted effort to incorporate the ability to test many of these fastidious organisms in their test systems. It is probably safe to assume that the instrument manufacturers will probably meet with increasing success in the use of these enhancements to their primary systems.

CONCLUSION

Only the most prominent issues have been included in this section. The disk diffusion method has certainly withstood the test of time and has survived in spite of the presence of many automated alternatives. It is difficult to perceive of an automated system that can compete with the flexibility and low cost of the disk diffusion method.

Although this issue section has focused on problems created by the automated test systems, it should be noted that significant progress has been made by all the system manufacturers in improving their systems. The combination of these automated systems with their ever increasing software enhancements, combined with continued integration with laboratory computer systems, offer a range of benefits that cannot be denied. This is the strength of these systems. The challenge of the '90's rests with the instrument manufacturers to address and resolve many of the problems outlined in this section with creative solutions. The talent that has been assembled in recent years by these manufacturers is a strong indication of how the field will progress in the future. Microbiologist outside the United States often accuse their U.S. counterparts of being over-fixated on automation for susceptibility testing. This may be true, but complex technological developments always create some problems and solve many others. Most marketing experts predict significant increases in automation in the rest of the industrialized world. The accent for the '90's will depend on minimizing the problems, and finding creative solutions. The concept of employing these systems is now well established, and their use will most likely increase in spite of pressure to reduce medical costs in the coming years. The remainder of this century should be an exciting time for both the clinical microbiologist and the instrument manufactures.

THE USE OF IN-VITRO KINETIC MODELS IN THE EVALUATION OF ß-LACTAM/ß-LACTAMASE INHIBITOR COMBINATIONS

Christine E. Thorburn and Brian Slocombe

SmithKline Beecham Pharmaceuticals
Brockham Park
Betchworth, Surrey, RH37AJ

INTRODUCTION

In conventional *in vitro* tests, bacteria are exposed to a constant concentration of antibiotic throughout the period of the test, and in the case of ß-lactam/ß-lactamase inhibitor combinations the concentrations of each component will be fixed. This does not reflect the situation in man, where the concentrations of antibiotic and inhibitor in the serum and tissues will be constantly changing according to absorption and elimination rates and any metabolism of the compounds. *In vitro* kinetic models can be used to simulate the concentrations of antibiotics measured in the serum and extravascular fluid of man following conventional dosage and to assess their antibacterial activities. In these studies, *in vitro* kinetic models have been used in various ways to assess the bactericidal activities of ß-lactam/ß-lactamase inhibitor combinations. Straightforward comparative studies can be carried out using different antibiotics; comparisons between different ß-lactams and ß-lactamase inhibitors; studies to look at concentrations of inhibitor required for different pathogens; and, interaction studies to look for synergy or antagonism can be carried out using these models. Examples of all these types of experiment are described below.

In Vitro Kinetic Model

The *in vitro* model used in these studies is essentially the open, one-compartment model originally described by Grasso in 1978[1] and comprises four glass vessels; a reservoir, containing sterile, drug-free nutrient broth; an antibiotic flask, A and a culture flask, B, which contain constant volumes of broth; and an overflow (Figure 1). The system is set up in a constant temperature room at 37° C. At the start of the experiment the culture flask contains a rapidly-growing bacterial culture, but no antibiotic. The antibiotic is injected into flask A, through a self-sealing port, and the pump is switched on. Drug-free broth medium is pumped from the reservoir into flask A, and antibiotic-containing broth is, in turn, forced under pressure into flask B, so that the concentration of antibiotic in flask A falls at a constant rate, whereas that in flask

Antimicrobial Susceptibility Testing, Edited by
J.A. Poupard *et al.*, Plenum Press, New York, 1994

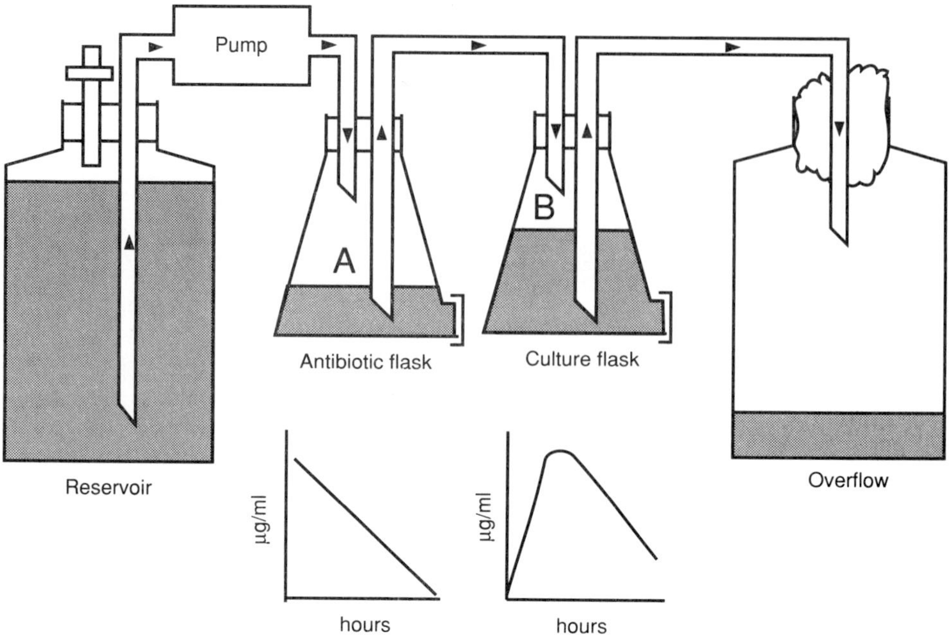

Figure 1. Schematic diagram of the open, one-compartment kinetic model.

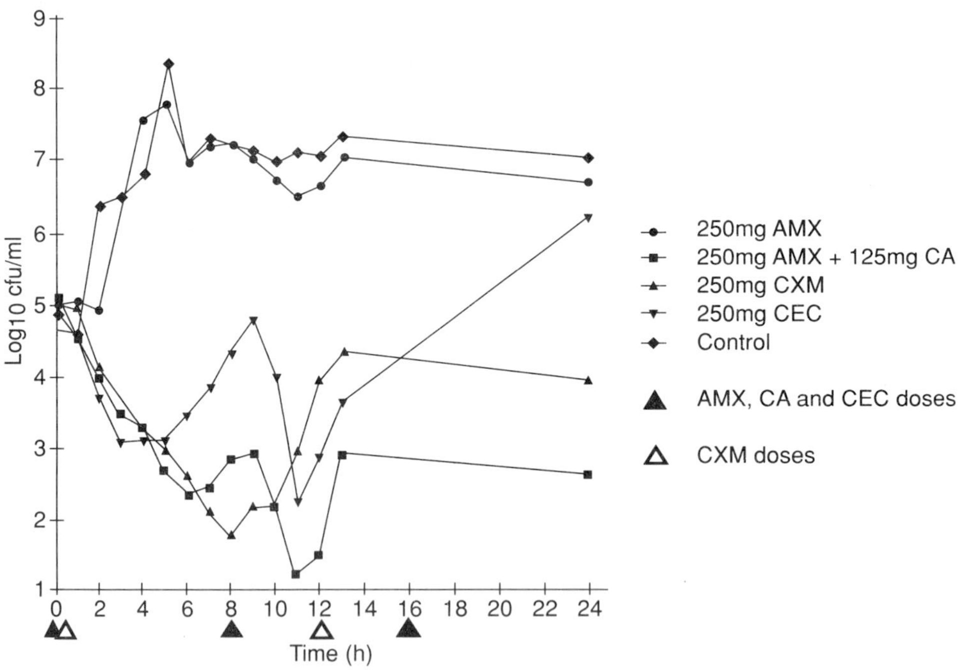

Figure 2. Bacterial activity of simulated serum concentrations of amoxycillin (AMX) plus clavulanic acid (CA), cefuroxime (CXM) and cefaclor (CEC) achieved following conventional oral dosage, against *S. aureus* NCTC 11561.

B (the culture flask) increases and then falls, simulating absorption and elimination rates seen in man. The rates at which the concentration rises and falls can be altered by changing the flow rate of the pump and the volumes in the flasks. Samples are taken from the culture flask in order to measure the concentration of antibiotic present (usually by microbiological assay) and to check that it coincides with the human data, as well as to count the number of viable bacteria present to determine the bactericidal effect. The antibiotic doses can be delivered to flask A via a pump interfaced to a computer, which allows the automatic administration of repeat doses and the simulation of different dosing regimens.

Comparative Studies with Oral ß-Lactam Antibiotics

Concentrations of amoxycillin and clavulanic acid achieved in the serum of man following oral dosage of 250 mg + 125 mg respectively were simulated in the model, and repeat doses were given 8-hourly for 24 hours. The bactericidal activity of these concentrations was compared with those of simulated serum concentrations following 250 mg amoxycillin used alone, 250 mg cefuroxime[2] dosed 12-hourly and 250 mg cefaclor[3] dosed 8-hourly against a ß-lactamase-producing strain of *Staphylococcus aureus*, NCTC 11561 (Figure 2). Although initially rapidly bactericidal, the culture treated with 250 mg cefaclor regrew from 5 hours and, despite further bactericidal effect following the 8-hourly doses, had almost fully regrown by 24 hours. Cefuroxime was more bactericidal than amoxycillin/clavulanic acid up to 8 hours, but 12-hourly cefuroxime was not as effective as amoxycillin/clavulanic acid dosed 8-hourly over 24 hours.

Synergy Studies with a Range of Concentrations of ß-Lactamase Inhibitor

Studies have also been performed in the model to determine whether concentrations of clavulanic acid achieved in the serum and tissues of man following oral dosage of 125 mg were sufficient to protect concurrent concentrations of amoxycillin from the ß-lactamases produced by *S. aureus*, *Haemophilus influenzae* and *Moraxella catarrhalis*. Blister fluid concentrations achieved in man following oral dosage of 250 mg amoxycillin plus 125 mg clavulanic acid were simulated and shown to be bactericidal with no regrowth over 24 hours, against ß-lactamase-producing strains of all three organisms, whereas amoxycillin alone was ineffective (data not shown). What is more, much lower concentrations of clavulanic acid than those achieved in serum following a conventional 125 mg dose; i.e., one quarter of these data (equivalent to a 31.25 mg dose) or a constant concentration of 0.12 μg/ml, were able to totally protect simulated serum concentrations of amoxycillin. The protected amoxycillin produced bactericidal activity against the ß-lactamase-producing strain of *S. aureus*, NCTC 11561, similar to that seen using amoxycillin alone against the ß-lactamase-negative isogenic variant of this strain.

Comparative Studies with Intravenous ß-Lactam/ß-Lactamase Inhibitor Combinations

In this study, the activity of simulated serum concentrations of ticarcillin and clavulanic acid achieved following a 30-minute intravenous infusion of 3 g plus 100 mg was compared with that of simulated serum concentrations following a 15-minute intravenous infusion of 2 g ampicillin plus 1 g sulbactam against *E. coli* NCTC 11560,

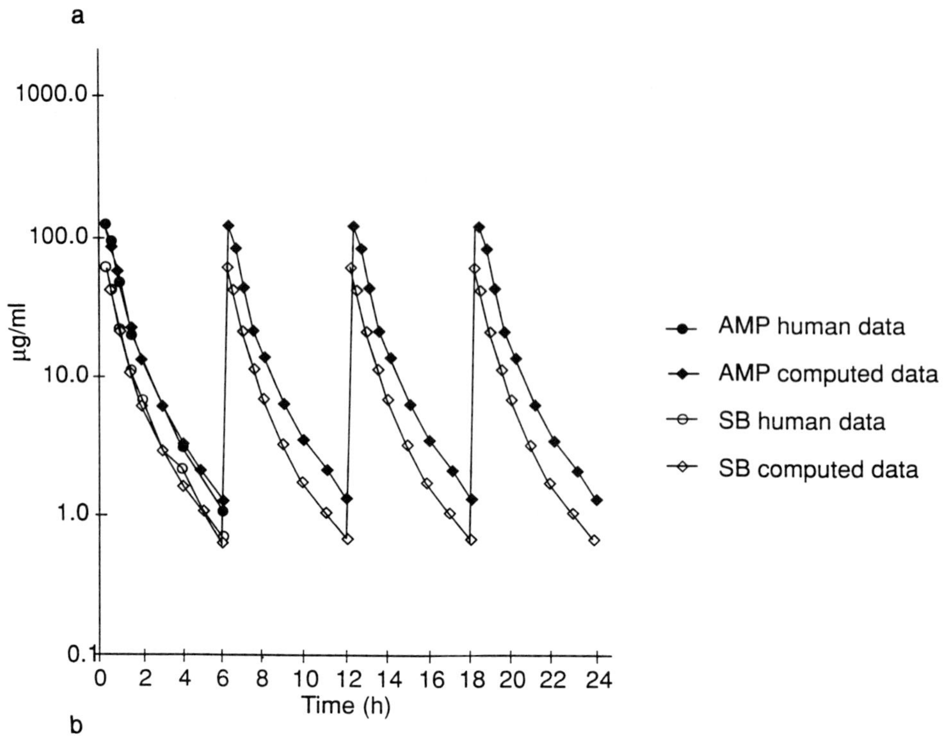

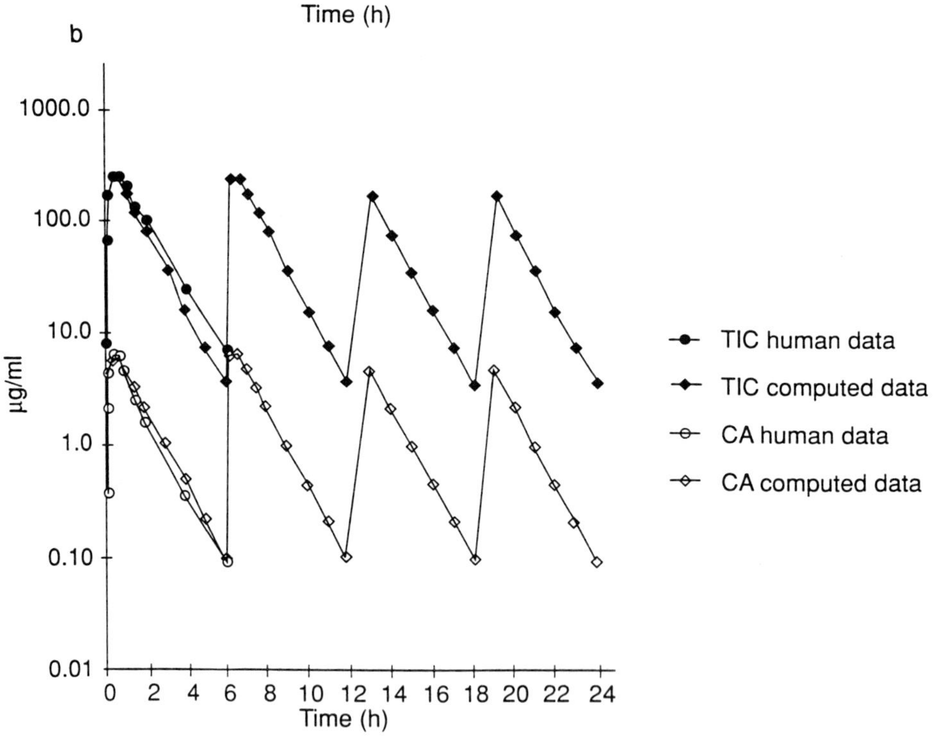

Figure 3. a) Concentrations of ampicillin (AMP) and sulbactam (SB) achieved in man following a 15-minute intravenous infusion of 2g AMP+1g SB and the predicted data for the model.
b) Concentrations of ticarcillin (TC) and clavulanic acid (CA) achieved in man following a 30-minute intravenous infusion of 3g TIC+100mg CA and the predicted data for the model.

which produces plasmid-mediated, TEM-1 ß-lactamase. In this experiment, the concentrations of ampicillin, sulbactam, ticarcillin and clavulanic acid were all close to the human data[4,5] (Figure 3), except where ampicillin was used alone. Here, the ß-lactamase present hydrolysed the ampicillin so that lower concentrations than predicted were measured from the first dose. Hydrolysis of ticarcillin alone was also seen, but only following the second dose. Consequently, ampicillin and ticarcillin were both ineffective when tested alone (Figure 4). The bactericidal activity of ticarcillin/clavulanic acid was greater than that of ampicillin/sulbactam and the culture treated with ampicillin/sulbactam had fully regrown, to 3.8×10^8 cfu/ml, by 24 hours, whereas that treated with ticarcillin/clavulanic acid contained 4.7×10^5 cfu/ml at this time. Moreover, after 24 hours exposure to ampicillin/sulbactam the cultures were 4-fold less susceptible to ampicillin/sulbactam and 2 to 16-fold less susceptible to ticarcillin/clavulanic acid, whereas the susceptibilities of the cultures exposed to ticarcillin/clavulanic acid for 24 hours were unchanged.

Investigation of Antagonistic Interactions

Antagonistic interactions between ticarcillin and clavulanic acid have occasionally been reported using the more conventional *in vitro* tests against organisms which produce inducible, chromosomally-mediated Class I ß-lactamases.[6] Two strains, *P. aeruginosa* TM0004 and *E. cloacae* T626, which show antagonism by disk and MIC tests respectively, were tested in the *in vitro* kinetic model, against simulated serum concentrations following a 30-minute intravenous infusion of 3 g ticarcillin alone, or in the presence of a simulated dose of 200 mg clavulanic acid. The higher dose of clavulanic acid was used to obtain maximum inducing potential. The results in the model showed that, despite antagonistic interactions between ticarcillin and clavulanic acid in conventional tests due to induction of ß-lactamase, no antagonism was apparent using concentrations of the agents which are measured in man, due to the high concentrations of ticarcillin achieved and its good stability to the ß-lactamases involved.

CONCLUSIONS

In vitro kinetic models can be used for a wide range of studies, some of which are described here. They can be used to simulate concentrations of antibiotics achieved in the serum and extravascular fluids of man following oral, intramuscular or intravenous dosage. As demonstrated by the data described above, the model may be used to study; direct comparative activity, the effect of different dosage regimens to predict breakpoint concentrations more accurately, interaction studies looking for either synergy[7] or antagonism at concentrations likely to be achieved in man and comparisons between different combinations of ß-lactams and ß-lactamase inhibitors. Another useful feature of this model is that long antibiotic exposure to continuously fluctuating concentrations can be examined. Although there are many factors which exist *in vivo* which are not taken into account in these experiments, it is hoped that the data produced are closer to the clinical situation than those achieved by using the more conventional *in vitro* tests.

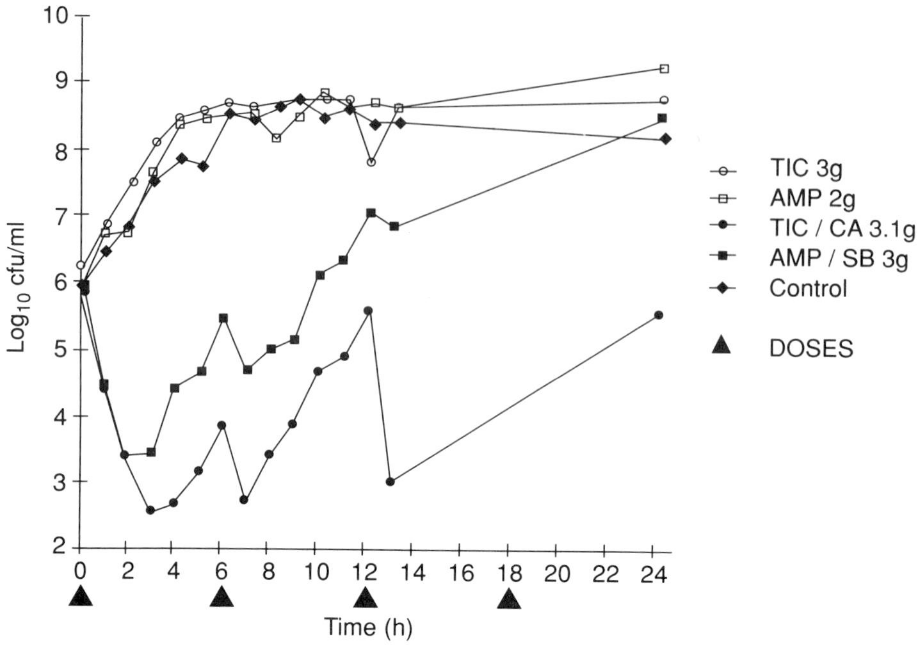

Figure 4. Bacterial activity of simulated serum concentrations of ampicillin plus sulbactam and ticarcillin plus clavulanic acid achieved following conventional intravenous dosage, against *E. Coli* NCTC 11560.

REFERENCES

1. S. Grasso, G. Meinardi, I. deCarneri, and V. Tamassia, New In Vitro Model to Study the Effect of Antibiotic Concentration and Rate of Elimination on Antibacterial Activity, Antimicrobial Agents and Chemother. 13:570-576 (1978).
2. S.M. Harding, P.E.O. Williams, and J. Ayrton, Pharmacology of Cefuroxime as the 1-Acetoxyethyl ester in Volunteers, Antimicrob. Agents and Chemother. 25, 1:78-82 (1984).
3. A. Glynne, R.A. Gouldbourn, and R. Ryden, A Human Pharmacology Study of Cefaclor, J. Antimicrob. Chemother. 4:343-348 (1978).
4. G. Foulds, Pharmacokinetics of Sulbactam and Ampicillin in Humans - A Review, Reviews of Infect. Diseases 8, 5:S503-511 (1986).
5. D.H. Staniforth, P.E. Coates, B.E. Davies, and R. Horton, Pharmacokinetics of Parenteral Ticarcillin Formulated with Clavulanic Acid - Timentin, Int. J. Clin. Pharm. Ther. and Toxic. 24, 3:123-129 (1986).
6. S.J. Cavaleiri, C.C. Sanders, and C. New, Influence of ß-Lactamase Inhibitors on the Potency of Their Companion Drug with Organisms Possessing Class I Enzymes, Antimicrob. Agents and Chemother. 35, 7:1343-1347 (1991).
7. A.R. White, D.H. Stokes, B. Slocombe, and R. Sutherland, Bactericidal Effects of Amoxycillin/Clavulanic Acid and Ticarcillin/Clavulanic Acid in *In Vitro* Kinetic Models, J. Antimicrob. Chemother. 15, A:227-232 (1985).

PREVALENCE OF TICARCILLIN/CLAVULANIC ACID-RESISTANT ENTEROBACTERIACAEAE IN NINE SEPARATE MEDICAL CENTERS DURING THE YEARS 1983, 1989, and 1991

Arthur L. Barry

The Clinical Microbiology Institute
Tualatin, Oregon

INTRODUCTION

The combination of ticarcillin with clavulanic acid has proven to be an effective antimicrobial agent. For *in vitro* susceptibility testing doubling dilutions of ticarcillin are combined with a fixed concentration of 2 μg clavulanic acid per ml. [5,6] Resistance to ticarcillin/clavulanic acid (T/C) has been relatively uncommon among the *Enterobacteriaceae*. [2,5,6,9] In 1988, Sanders et al. described strains of *Escherichia coli* and *Klebsiella pneumoniae* that were T/C resistant because they produced excess amounts of ß-lactamase enzymes which could not be completely neutralized by the low concentration of clavulanic acid (2 μg/ml) that is normally tested *in vitro*. [11] Whether strains with elevated T/C MICs should be considered clinically resistant is a matter of presumption. The inability of the disk diffusion test to detect such hyper-producers of ß-lactamase enzymes has been studied, but the clinical relevance of those discrepant results has yet to be evaluated. [3,7,11]

Because T/C has been widely used for a good many years, it seems appropriate to determine whether resistance to T/C has changed over time. Our first survey was carried out in 1983 and published in 1984[2] and two subsequent studies were performed in 1989[9] and again in 1991[3]. This report re-examines results of all three surveys in order to determine whether there has been a change in the prevalence of T/C resistant strains over the years. Specifically, T/C resistance among *Escherichia coli* and *Klebsiella pneumoniae* was documented to estimate the prevalence of ß-lactamase hyperproducing strains in different medical centers and in different years. In addition, previously unpublished data compared T/C to ampicillin alone as well as ampicillin/sulbactum and amoxicillin/clavulanic acid.

In 1983, 1989 and 1991, only 7% to 8% of enteric bacilli were resistant to T/C and another 7% to 10% were moderately susceptible. There was no evidence that the prevalence of resistant strains has changed over the eight year span of time, but there was variation from center to center. Among *Escherichia coli*, resistance to T/C occurred at rates ranging from 3% to 12% in different centers and 0% to 13% of *K. pneumoniae* were resistant to T/C. In different centers 2% to 25% of *S. aureus* isolates and 24% to 56% of coagulase-negative staphylococci were resistant to T/C (MIC ≥ 8.0/2.0 μg/ml).

MATERIALS AND METHODS

Three different studies were coordinated in 1983, 1989 and in 1991 in order to document the *in vitro* activity of T/C against consecutively isolated bacterial species in different medical centers. Results of the 1983 evaluation have been published previously[2] and are included in this report to determine whether there has been a change in the prevalence of T/C-resistance in subsequent studies. Portions of the other evaluations have also been published in a different context[3,9], but this report only includes the broth microdilution test results that can be compared to those obtained previously.

For each study, 3 to 5 different participating laboratories performed broth microdilution tests against all bacterial isolates that were considered significant enough to be selected for routine susceptibility tests. This sampling process was continued for 30-45 days or until a predetermined number of isolates was tested. Broth microdilution tests were performed exactly as described by the National Committee for Clinical Laboratory Standards.[10] Quality control strains and other reference strains were tested as the data were collected and the results of those tests documented an acceptable degree of inter-laboratory and intra-laboratory precision and accuracy.

The laboratories that participated in the three different studies are described in Table 1. Only one participant was involved in all three studies (laboratory C) and laboratories D and E were involved in two of the three studies. A review of these data documented the prevalence of T/C resistance in nine different medical centers.

RESULTS

T/C activity over time

Table 2 describes the percentage of strains inhibited by different concentrations of ticarcillin with 2 μg clavulanic acid per ml. In previous years, gram-negative bacilli that were inhibited by (≤ 64/2 μg/ml) were considered susceptible to the T/C combination. Recently, interpretive criteria have been modified to identify two categories, ie., Susceptible (MIC ≤ 16/2 μg/ml) and Moderately Susceptible (MIC 32/2 or 64/2 μg/ml). The cumulative percentage data in Table 2 allows the reader to apply different interpretive criteria to the T/C MIC data. Species represented by relatively small numbers were omitted from this analysis. In the 1991 study, only the *Enterobacteriaceae* and *Pseudomonas aeruginosa* were tested. Between 1983 and 1991, there was no substantial change in the overall percentage of enteric bacilli inhibited by T/C at ≤ 16/2 μg/ml or at ≤ 64/2 μg/ml.

T/C Resistance in Different Medical Centers

Table 3 presents the percentage of enteric bacilli that were moderately

Table 1. Participating Facilities and Laboratory Director Contributing to Three Separate Studies which Documented Susceptibility of Consecutively Isolated Bacterial Pathogens to Ticarcillin/Clavulanic Acid.

Laboratory Code and Years of Study	Participating Laboratory Director - Facility and Location
A 1983	T.L. Gavan - The Cleveland Clinic Foundation, Cleveland, OH
B 1983	L.W. Ayers - The Ohio State University Medical Center, Columbus, OH
C 1983, 1989, 1991	E.H. Gerlach - St. Francis Medical Center, Wichita, KS
D 1989, 1991	P.C. Fuchs - St. Vincent Hospital and Medical Center, Portland, OR
E 1989, 1991	M.A. Pfaller - University of Iowa Hospitals and Clinics, Iowa City, IA
F 1989	S.D. Allen - Indiana University Medical Center, Indianapolis, IN
G 1989	K. Aldridge - Louisiana State University Medical Center, New Orleans, LA
H 1991	J.C. McLaughlin - University of New Mexico Medical Center, Albuquerque, NM
I 1991	D.J. Hardy - University of Rochester Medical Center, Rochester, NY

Table 2. *In vitro* Activity of Ticarcillin with 2.0 ug Clavulanic Acid per ml Against Common Bacterial Pathogens in Three Separate Multi-Laboratory Studies.

Microbial Species and study Year	Number of Isolates	Cumulative % Inhibited (ug/ml)[a] ≤ 4.0	8.0	16	32	64
Escherichia coli						
1983	1,875	76	81	84	88	93
1989	1,375	68	75	81	87	94
1991	1,335	71	78	83	89	94
Citrobacter diversus						
1983	73	73	84	92	97	100
1989	87	86	94	95	97	97
1991	20	100	-	-	-	-
Citrobacter freundii						
1983	199	75	76	78	79	86
1989	60	68	70	70	75	80
1991	69	59	64	67	68	78
Enterobacter aerogenes						
1983	259	65	67	70	74	93
1989	102	67	72	74	82	90
1991	73	60	69	69	73	81
Enterobacter cloacae						
1983	516	64	68	72	75	79
1989	228	44	50	55	59	71
1991	154	55	64	67	71	78
Klebsiella oxytoca						
1983	181	60	74	85	90	92
1989	91	87	92	95	97	97
1991	79	87	94	95	96	98

Microbial Species and study Year	Number of Isolates	Cumulative % Inhibited (ug/ml)[a] ≤ 4.0	8.0	16	32	64
Klebsiella pneumoniae						
1983	908	63	77	86	90	93
1989	438	78	87	91	95	97
1991	365	81	87	91	94	96
Serratia marcescens						
1983	311	62	75	84	87	90
1989	122	58	76	85	93	96
1991	64	44	69	89	95	100
Proteus mirabilis						
1983	617	99	99	100	-	-
1989	277	98	99	99	99	99
1991	161	100	-	-	-	-
Proteus vulgaris						
1983	49	98	100	-	-	-
1989	29	97	100	-	-	-
1991	18	94	100	-	-	-
Morganella morganii						
1983	165	76	87	93	96	99
1989	49	78	90	94	98	100
1991	23	52	70	96	96	96
Providencia rettgeri						
1983	21	81	95	100	-	-
1989	6	100	-	-	-	-
1991	7	71	71	100	-	-
Providencia stuartii						
1983	59	78	81	86	92	97
1989	5	100	-	-	-	-
1991	8	100	-	-	-	-

(Continued)

Table 2. (Continued)

Microbial Species and study Year	Number of Isolates	Cumulative % Inhibited (ug/ml)[a] ≤ 4.0	8.0	16	32	64
Pseudomonas aeruginosa						
1983	1,385	4	12	60	77	88
1989	578	5	16	50	70	83
1991	253	5	14	57	81	91
Xanthomonas maltophilia[b]						
1983	64	38	55	69	75	84
1989	89	30	46	61	75	87
Acinetobacter calcoaceticus[b]						
1983	148	48	76	97	98	98
1989	95	54	75	90	95	97
Staphylococcus aureus[b]						
1983	1,137	95[c]	95	100	-	-
1989	821	90[c]	93	96	98	99
Coagulase-Negative Staphylococci[b]						
1983	689	68[c]	79	86	90	92
1989	675	53[c]	66	76	81	85
Enterococcus faecalis[b]						
1983	1,100	1	1	4	37	95
1989	394	0	1	3	18	86

[a] Inhibitory concentration of ticarcillin in the presence of clavulanic acid (2.0 ug/ml).

[b] The 1991 study was limited to the *Enterobacteriaceae* and *Pseudomonas aeruginosa*.

[c] *Staphylococci* inhibited by ≤4.0/2.0 ug/ml are considered susceptible to ticarcillin/clavulanic acid (1).

susceptible to T/C (MIC 32/2 or 64/2 μg/ml) and the percentage that were resistant to T/C (MIC ≥ 128/2 μg/ml). These data were calculated for each medical center and for each of the three separate study years. This permitted an overview of the amount of variation from year to year and from center to center. In different years, T/C resistance occurred with 7% to 8% of all enteric bacilli and an additional 7% to 10% were moderately susceptible (83% to 85% were fully susceptible to T/C). In any one center, 6% to 10% of enteric bacilli were resistant to T/C and 78% to 88% were fully susceptible.

There has been recent interest in the prevalence of T/C resistant *E. coli* and *K. pneumoniae* because such strains are resistant by virtue of their ability to produce high levels of ß-lactamase enzymes which overwhelm the clavulanic acid *in vitro*. Over the years, 6% of *E. coli* have been resistant to T/C but that ranged from 3% to 12% in different medical centers and different years. Another 7% to 18% of *E. coli* were moderately susceptible to T/C (MIC 32/2 or 64/2 μg/ml). Overall, 73% to 90% of *E. coli* were susceptible to 16/2 μg of T/C per ml. Among the *K. pneumoniae* isolates, 77% to 97% were susceptible to T/C and 0% to 13% were resistant (MIC ≥ 128/2 μg/ml). Variation between institutions represents the greatest variable and that observation cannot be explained by differences in testing procedures since extensive controls confirmed methodological consistency in all testing facilities. Year-to-year variations in the overall percentage of T/C resistant strains is well within the range of variation that might be expected in samples of this nature.

Gram-Positive Cocci

The anti-staphylococcal activity of the T/C combination was documented in 1983 and again in 1989 (Table 4). Penicillinase-producing staphylococci would be expected to be resistant to ticarcillin alone but susceptible to the T/C combination. On the other hand, methicillin-resistant staphylococci should be expected to be resistant to the T/C combination. Barry et al.[1] recommended that T/C-susceptible strains should be defined as those with MICs ≤4.0/2.0 μg/ml, whereas methicillin-resistant staphylococci tend to have T/C MICs ≥ 8.0/2.0 μg/ml. The two surveys that included staphylococci did not document methicillin-susceptibility or resistance for each strain. However, we could separate staphylococci by their resistance to the T/C combination (MIC ≥8.0/2.0 μg/ml). Overall, 8% of *S. aureus* and 39% of the coagulase-negative staphylococci were resistant to T/C and were presumably methicillin- and oxacillin-resistant strains. There was a marked degree of variation between medical centers. Among *S. aureus* isolates, susceptibility to the T/C combination ranged from 75% to 98% in different centers. Among coagulase-negative staphylococci, susceptibility to T/C was much less prevalent (61%) and that ranged from 44% to 76%. We concluded that T/C continues to have significant antistaphylococcal activity in those institutions where methicillin-resistant strains occur infrequently. Nearly all ß-lactams are assumed to be inactive against methicillin-resistant staphylococci; the T/C combination is not unique in that respect.

Enterococcus faecalis is the only other gram-positive coccus that is included in this report. T/C, like ticarcillin, has little activity against *E. faecalis* or other enterococci (Table 2). Although some strains had MICs in the moderately susceptible category, other penicillins should have much better activity against the enterococci.[4]

Table 3. Pervalence of Enteric Bacilli Resistant of Moderately Susceptible to Ticarcellin/ Clavulanic Acid in Nine Different Medical Centers and Three Separate Surveys (1983, 1989 and 1991).

Laboratory Code	Year of Study	All Enterobacteriaceae No. tested	%MS[a]	%R[a]	*E. coli* only No. tested	%MS	%R	*K. pneumoniae* only No. tested	%MS	%R
A	1983	1,492	8	10	526	10	12	247	10	12
B	1983	1,721	9	6	563	11	4	294	8	4
C	1983	2,020	5	7	785	6	6	367	5	7
	1989	601	8	5	267	13	3	110	4	1
	1991	450	7	6	203	12	9	76	1	3
D	1989	415	11	6	247	12	5	54	11	0
	1991	486	8	5	339	9	5	56	9	2
E	1989	546	14	8	202	18	9	83	6	0
	1991	502	8	7	253	9	6	69	3	0
F	1989	566	6	9	279	7	3	84	4	9
G	1989	741	10	6	385	14	9	107	8	4
H	1991	481	10	7	286	13	5	95	1	4
I	1991	457	11	9	254	13	6	69	10	13
ALL	1983	5,233	7	8	1,875	9	7	908	7	7
ALL	1989	2,869	10	7	1,375	13	6	438	6	3
ALL	1991	2,376	9	7	1,335	11	6	365	5	4

[a]Percentage of strains Moderately Susceptible (MS, MIC 32/2 or 64/2 μg/ml) and percentage Resistant (R, MIC ≥128/2 μg/ml), all others were fully susceptible.

T/C versus other Antimicrobial Agents

To put these T/C data into perspective, the 1989 study included tests with two other ß-lactamase inhibitor combinations as well as ampicillin alone. Table 5 describes the percentage of strains that were fully susceptible to the four different drugs. Among all *Enterobacteriaceae*, 35% to 53% of strains were susceptible to ampicillin with sulbactam (2:1 ratio). A 2:1 combination of amoxicillin with clavulanic acid, on the other hand, was effective against 62% to 77% of all isolates. Ticarcillin with clavulanic acid was even more active, ie., 84% to 87% of all enterics were susceptible.

Among the 1,375 strains of *E. coli*, 67% were susceptible to ampicillin alone (58% to 77% in different medical center). With the addition of sulbactam (2:1 ratio), ampicillin inhibited 73% of the *E. coli* (64% to 80% in different centers). Amoxicillin/clavulanic acid and ticarcillin/clavulanic acid combinations both inhibited 81% of all *E. coli*.

As expected, ampicillin by itself had very little activity against *K. pneumoniae* isolates, but 81% (74% to 96%) of these strains were susceptible to the ampicillin/sulbactam combination. The two clavulanic acid combinations were more effective since 91% (86% to 99%) of all strains were susceptible to one or both of the combinations.

As expected, ampicillin, ampicillin/sulbactam and amoxicillin/clavulanic acid had no activity against *Pseudomonas aeruginosa*, but ticarcillin did have antipseudomonal activity. In different medical centers 68% to 96% of *P. aeruginosa* isolates were susceptible to ticarcillin/clavulanic acid.

DISCUSSION

The results of the three multi-laboratories studies that are reviewed in this report lead one to conclude that there is no clear-cut tendency for increased resistance to ticarcillin/clavulanic acid over the eight-year span of time covered. There were, however, important differences in T/C susceptibility patterns in different medical centers at any given time. It is most likely that T/C resistant strains may be endemic in one or more areas within a given institution and the type of microorganisms that are endemic in each area should be expected to change from time to time. Because of the complex sources of the isolates being tested during the short period of time that was being sampled in different medical centers, it is not surprising that there was some variation from center to center. In all three studies, the sample consisted of all isolates selected for susceptibility tests over a defined period of time or until a predetermined number of strains was tested. That provided an unbiased sample of isolates that were being encountered in that period of time. The most common species (*E. coli*) predominates in such a sample. No attempt was made to exclude multiple isolates from one patient, multiple isolates from different body sites or epidemiologically related patients infected with the same endemic strain. In a survey of this nature, it is not possible to subdivide the isolates by patients, body sites, or hospital locations. The data that are presented simply document the true prevalence of T/C resistance in different institutions and at different times.

In 1988 Sanders et al.[11] described strains of piperacillin-resistant *E. coli* and *K. pneumoniae* that were also resistant to ticarcillin/clavulanic acid. Some of those

Table 4. Antistaphylococcal Activity of Ticarcillin/Clavulanic Acid Against Isolates Recovered from Seven Different Medical Centers in Two Separate Surveys (1983 and 1989).

Laboratory Code	Year of Study	*S. aureus* No. tested	*S. aureus* % Resist.[a]	Coag. Neg. Staph.[b] No. tested	Coag. Neg. Staph.[b] % Resist.[a]
A	1983	559	8%	430	24%
B	1983	207	2%	143	47%
C	1983	372	3%	56	45%
C	1989	197	4%	80	54%
D	1989	135	6%	104	27%
E	1989	181	25%	256	50%
F	1989	146	16%	154	46%
G	1989	162	3%	82	56%
ALL	Both	1,958	8%	1,304	39%

[a]Staphylococci with ticarcillin/clavulanic acid MICs ≥8.0/2.0 μg/ml were categorized as being resistant (1).

[b]Coagulase-negative *Staphylococcus* species, predominantly *S. epidermidis*.

Table 5. Perecentage of Strains Susceptible to Each of Four Antimicrobial Agents in Five Different Medical Centers[a] Surveyed in 1989.

Microorganism and Antimicrobial Agent	Susceptible MIC (μg/ml)	% of Susceptible[b] Strains in each medical center					
		C	D	E	F	G	Combined
All *Enterobacteriaceae* (2,869 strains)							
Ampicillin	≤8.0	45	53	35	40	46	44
Ampicillin/sulbactam	≤8.0/4	74	70	57	69	65	67
Amoxicillin/clavulanic acid	≤8.0/4	77	72	62	72	67	70
Ticarcillin/clavulanic acid	≤16/2	87	83	78	85	84	83
Escherichia coli (1,375 strains)							
Ampicillin	≤8.0	77	74	65	62	58	67
Ampicillin/sulbactam	≤8.0/4	80	76	70	80	64	73
Amoxicillin/clavulanic acid	≤8.0/4	88	81	83	90	66	81
Ticarcillin/clavulanic acid	≤16/2	84	83	73	90	76	81

(continued)

Table 5. (Continued)

Microorganism and Antimicrobial Agent	Susceptible MIC (μg/ml)	% of Susceptible[b] Strains in each medical center C	D	E	F	G	Combined
Klebsiella pneumoniae (438 strains)							
Ampicillin	≤8.0	0	3	11	4	12	6
Ampicillin/sulbactam	≤8.0/4	90	78	89	74	75	81
Amoxicillin/clavulanic acid	≤8.0/4	97	87	99	86	86	91
Ticarcillin/clavulanic acid	≤16/2	96	89	94	87	88	91
Pseudomonas aeruginosa (578)							
Ampicillin	≤8.0	<1	0	<1	0	<1	<1
Ampicillin/sulbactam	≤8.0/4	<1	0	2	0	<1	<1
Amoxicillin/clavulanic acid	≤8.0/4	<1	0	2	0	<1	2
Ticarcillin/clavulanic acid	≤64/2	87	96	80	85	68	83

[a]Collaborating facilities are identified in Table 1.

[b]Excludes strains that are moderately susceptible or resistant to each drug.

isolates had T/C MICs in the susceptible (MIC ≤16/2 μg/ml) or moderately susceptible (MIC 32/2 or 64/2 μg/ml) whereas most of the strains were fully resistant (T/C MICs ≥128/2 μg/ml). Although Sanders et al.[11] described the T/C resistant strains in 1988, our results suggest that such strains existed as early as 1983 and the prevalence rates have not changed over the years. The overall prevalence of T/C resistant *E. coli* held steady at 6% to 7% but in different medical centers the prevalence rates ranged from 3% to 12% at different times. T/C resistant *K. pneumoniae* occurred less frequently (0% to 13% in different centers). The data included in this report provide no evidence that T/C resistance among *E. coli* or *K. pneumoniae* isolates has changed substantially since 1983.

Additional data presented in Table 4 document the extreme variability in the proportion of T/C-resistant staphylococci in different medical centers. In 1987-1988 a national survey documented the prevalence of oxacillin-resistant *S. aureus* among 40 different medical centers.[8] Oxacillin-resistance rates ranged from 0% to 42% in different centers. A similar degree of variation in T/C resistance in different medical centers should be expected. T/C does appear to be active against oxacillin-susceptible staphylococci and that activity might be useful in some clinical settings.

In vitro activity of T/C was compared to those of the two other ß-lactamase-inhibitor combinations that are currently being marketed. T/C was more active than ampicillin/sulbactam or amoxicillin/clavulanic acid against enteric bacilli, including *E. coli* and *K. pneumoniae.* Ticarcillin had a further advantage in that it had anti-pseudomonal activity which should not be influenced by the presence of a ß-lactamase inhibitor.[2,5,6] In summary, ticarcillin/clavulanic acid remains effective against a wide range of bacterial pathogens and bacterial resistance has not changed over the years.

REFERENCES

1. A.L. Barry, Antistaphylococcal activity of amoxicillin and ticarcillin when combined with clavulanic acid, evaluation of oxacillin-resistant and oxacillin-susceptible isolates. Diagn. Microbiol. Infect. Dis. 13:357-361 (1990).
2. A.L. Barry, L.W. Ayers, T.L. Gavan, E.H. Gerlach, and R.N. Jones, In vitro activity of ticarcillin plus clavulanic acid against bacteria isolated in three centers, Eur. J. Clin. Microbiol., 3:203-206 (1984).
3. A.L. Barry, P.C. Fuchs, E.H. Gerlach, D.J. Hardy, J.C. McLaughlin, and M.A. Pfaller, Ticarcillin and ticarcillin/clavulanic acid susceptibility tests: error rates for disk tests with consecutively isolated *Enterbacteriaceae*, Antimicrob. Agents Chemother., 36:137-143 (1992).
4. A.L. Barry, C. Thornsberry, R.N. Jones, and T.L. Gavan, In vitro activity of mezlocillin and azlocillin compared to that of four other penicillins and two aminoglycosides, Cleveland Clinic Quarterly 47:311-319 (1980).
5. P.C. Fuchs, A.L. Barry, and R.N. Jones, In vitro activity and disk susceptibility of Timentin: current status, Amer. J. Med. 79Suppl 5B:25-32 (1985).
6. P.C. Fuchs, A.L. Barry, C. Thornsberry, and R.N. Jones, In vitro activity of ticarcillin plus clavulanic acid against 632 clinical isolates, Antimicro. Agents Chemother., 25:392-394 (1984).
7. P.C. Fuchs, R.N. Jones, and A.L. Barry, Reassessment of susceptibility test interpretive criteria for ticarcillin and ticarcillin-clavulanic acid, J. Clin. Microbiol., 27:2475-2481 (1989).

8. R.N. Jones, A.L. Barry, R.V. Gardiner, and R.R. Packer, The prevalence of staphylococcal resistance to penicillinase-resistant penicillin: a retrospective and prospective national surveillance trial of isolates from 40 medical centers, Diagn. Microbiol. Infect. Dis., 12:385-394 (1989).
9. R.N. Jones, M.A. Pfaller, P.C. Fuchs, K. Aldridge, S.D. Allen, and E.H. Gerlach, Piperacillin/Tazobactam (YTR830) combination, comparative antimicrobial activity against 5889 recent aerobic clinical isolates and 60 *Bacteroides fragilis* group strains, Diagn. Microbiol. Infect. Dis., 12:489-494 (1989).
10. National Committee for Clinical Laboratory Standards, Methods for dilution antimicrobial susceptibility tests for bacteria that grow aerobically, Approved Standard M7-A2 (2nd ed.) NCCLS, Villanova, PA (1990).
11. C.C. Sanders, J.P. Jaconis, G.P. Bodey, and G. Samonis, Resistance to ticarcillin-potassium clavulanate among clinical isolates of the family *Enterobacteriacae*: role of PSE-1 ß-lactamase and high levels of TEM-1 and SHV-1 and problems with false susceptibility in disk diffusion tests, Antimicrob. Agents Chemother., 32:1365-1369 (1988).

CORRELATION OF MINIMUM INHIBITORY CONCENTRATION RESULTS BETWEEN THE VITEK SYSTEM AND THE BIOMIC SYSTEM

Tammy L. Wolfram,[1] C. Ross McFarland,[2] and James A. Poupard[3]

St. Joseph Hospital, Reading, PA[1]
Pottstown Memorial Medical Center, Pottstown, PA[2]
SmithKline Beecham, King of Prussia, PA[3]

INTRODUCTION

This study compared the BIOMIC system to the Vitek system to determine the percentage of agreement between the minimum inhibitory concentrations. The BIOMIC semi-automated system employs disk diffusion test zone diameters to determine minimum inhibitory concentrations. Results were in agreement when the BIOMIC minimum inhibitory concentration was within ± 1 doubling dilution of the Vitek result. A total of 137 clinical isolates, including *Enterobacteriaceae*, non-fermenters, staphylococci, and enterococci were tested. The overall agreement for 1335 organism/drug combinations was 85%, with 10% minor and 5% major discrepancies. The agreement by organism group was 87% for *Enterobacteriaceae*, 87% for nonfermenters, and 79% for enterococci and staphylococci. Antimicrobic combinations that exhibited a high discrepancy rate included tetracycline vs. *Enterobacteriaceae* (39%) and amikacin vs. nonfermenters (30%). An agreement of 100% was found for ciprofloxacin and imipenem vs. nonfermenters, as well as for clindamycin and oxacillin vs. staphylococci. This study showed the BIOMIC system to be an acceptable, cost-effective alternative for determining minimum inhibitory concentrations.

The use of automated microdilution systems for minimum inhibitory concentration (MIC) testing has become increasingly popular during the past decade. Since cost-containment has become a critical issue in the clinical laboratory, the selection of a less costly MIC testing system would be advantageous.

In 1985 the BIOMIC antimicrobial susceptibility test system was developed by Giles Scientific, Inc, New York, NY to generate MICs, inhibitory quotients, and qualitative breakpoint categories from disk diffusion test zone diameters. The BIOMIC system calculates MIC values by direct regression analysis with the system algorithm on a continuous scale. The inverse linear relationship between zone diameters and MICs was established by Ericsson and Sherris in 1971.[6]

In addition to the relatively low instrument purchase price, there are several other advantages of MIC testing with the BIOMIC system. A decreased materials cost/test is realized due to the lower cost of disk diffusion test materials in comparison to automated microdilution panels. The flexibility of the method allows for additional cost savings, since the testing of undesired antimicrobics is avoided. The high accuracy and reproducibility of the disk diffusion method are other advantages of BIOMIC testing. Also, MIC testing of fastidious organisms is possible with the BIOMIC system.

The purpose of this study was to compare the BIOMIC system to a highly utilized automated system, the Vitek system (Vitek Systems, Hazelwood, MO) to determine the percentage of agreement between results.

MATERIALS AND METHODS

Bacterial isolates A total of 137 clinical isolates were obtained from 4 sites: St. Joseph Hospital, Reading, PA (106 isolates); Pottstown Memorial Medical Center, Pottstown, PA (21 isolates); Lehigh Valley Hospital Center, Allentown, PA (8 isolates); and Quality Control Medical Laboratories, Reading, PA (2 isolates). Only one isolate of a particular organism type was tested per patient. Table 1 lists the number and types of organisms tested.

Antimicrobial selection Five different antimicrobial batteries were selected. The antimicrobic batteries are found in Table 2.

BIOMIC testing The BIOMIC system consists of a software program that is run in an IBM-compatible personal computer. A printer was not utilized for the study, but must be purchased separately if desired. An electronic caliper is provided with the system for the purpose of measuring and subsequently entering zone diameters. Also included with the system is a plate-caliper stand, designed to provide a flat background at an optimal reading angle.

Disk diffusion testing was performed in accordance with the recommendations of the National Committee for Clinical Laboratory Standards.[8] Mueller Hinton broth (Remel, Lenexa, KS) was utilized for the incubation and inoculation of isolates. After 2-4 h of incubation at 37°C, the turbidity of the broth was adjusted to a 0.5 McFarland standard. Adjusted broth suspensions of the organisms were swabbed confluently onto 150 x 15 mm Mueller Hinton plates (Remel). Antimicrobic disks (Difco Laboratories, Detroit, MI) were then applied with a dispenser. Plates for *Staphylococcus* spp. isolates were incubated in a 35°C incubator for 24 h. All other plates were incubated at 37°C for 16-18 h. After the incubation period, zone diameters were recorded using the BIOMIC system.

Vitek testing The following Vitek susceptibility cards were used: GNS-F1, GNS-F2, GNS-T, GNS-UA, GPS-SA, and GPS-TA. Testing was performed according to Vitek system procedures.

Control strains Quality control was performed as suggested by the National Committee for Clinical Laboratory Standards. The following American Type Culture Collection strains were utilized: *Enterococcus faecalis* 29212, *Esherichia coli* 25922, *E. coli* 35218, *Pseudomonas aeruginosa* 27853, *Staphylococcus aureus* 25923, and *S. aureus* 29213.

Table 1. Number and Type of Organisms

Organisms	No. of clinical isolates[a]
Acinetobacter calcoaceticus biotype *anitratus*	8
Citrobacter diversus	4
Citrobacter freundii	6
Coagulase-negative staphylococci	12
Enterobacter aerogenes	6
Enterobacter agglomerans	1
Enterobacter cloacae	10
Enterobacter gergoviae	1
Enterococcus faecalis	10
Escherichia coli	11
Klebsiella oxytoca	2
Klebsiella pneumoniae	8
Proteus mirabilis	10
Pseudomonas aeruginosa	25
Serratia liquefacians	1
Serratia marcescens	11
Staphylococcus aureus	11

[a] A total of 137 clinical isolates were tested.

Table 2. Antimicrobic Batteries

Enterobacteriaceae non-urinary isolates	*Enterobacteriaceae* urinary isolates	*Enterococcus faecalis*	*Pseudomonas aeruginosa* and *Acinetobacter calcoaceticus* biotype *anitratus*	*Staphylococcus* spp.
Amikacin	Ampicillin	Ampicillin	Amikacin	Ampicillin/sulbactam
Ampicillin	Cefazolin	Erythromycin	Aztreonam	Cephalothin
Cefamandole	Cefoxitin	Nitrofurantoin	Carbenicillin	Ciprofloxacin
Cefoxitin	Gentamicin	Norfloxacin	Ceftazidime	Clindamycin
Cephalothin	Nitrofurantoin	Penicillin G	Ceftriaxone	Erythromycin
Gentamicin	Norfloxacin	Tetracycline	Ciprofloxacin	Oxacillin
Tetracycline	Tetracycline	Vancomycin	Gentamicin	Penicillin G
Ticarcillin	Ticarcillin		Imipenem	Tetracycline
Trimethoprim/sulfamethoxazole	Trimethoprim/sulfamethoxazole		Piperacillin	Trimethoprim/sulfamethoxazole
			Tetracycline	Vancomycin
			Ticarcillin/clavulanic acid	
			Trimethoprim/sulfamethoxazole	

Data analysis BIOMIC MICs were considered in agreement with the Vitek MICs when they were within ±1 doubling dilution. Because the Vitek MICs were frequently reported as less than or equal to, or greater than or equal to, BIOMIC MICs were considered in agreement when they were above or below the Vitek MIC as indicated.

Discrepancies were classified as major or minor. A major discrepancy occurred when the BIOMIC MIC was more than 1 doubling dilution different from the Vitek MIC, resulting in any change in the National Committee for Clinical Laboratory Standards breakpoint category call. Minor discrepancies occurred when BIOMIC results differed by more than 1 doubling dilution without a change in category call.

RESULTS

This study showed an 85% overall agreement between the BIOMIC system and Vitek system for 1335 organism/drug combinations. The percentage of minor and major discrepancies was 10% and 5%.

Table 3 shows the results for members of the family *Enterobacteriaceae*. The agreement ranged from 61% for tetracycline to 97% for cefazolin, gentamicin, and trimethoprim/sulfamethoxazole. The overall agreement between the two systems was 87% for 639 organism/drug combinations. Minor and major discrepancies represented 7% and 6% of the results, respectively. The highest percentage of minor discrepancies (29%) occurred for tetracycline. The trend was higher for tetracycline MICs calculated by the BIOMIC system. Nitrofurantoin testing resulted in the highest percentage of major discrepancies, 17%. In most cases, the Vitek MICs generated for nitrofurantoin were higher than BIOMIC MICs.

A summary of the results for nonfermentative gram-negative isolates is found in Table 4. The overall agreement for 396 organism/drug combinations was 87%, with 9% and 4% minor and major discrepancies. Agreement with amikacin was lowest at 70%. Discrepancies with amikacin were due to lower MICs calculated by the BIOMIC system. Tetracycline agreement was very low (12%) for *A. calcoaceticus* biotype *anitratus* as was found for the *Enterobacteriaceae*. Similar to the *Enterobacteriaceae* results, higher MICs were calculated by the BIOMIC for the nonfermenters. Two antimicrobics, ciprofloxacin and imipenem showed 100% agreement with the non-fermenters.

The overall agreement for *E. faecalis* and staphylococci was 79% for 300 organism/drug combinations. Results are shown in Table 5. The highest percentage of minor discrepancies occurred with *E. faecalis*. Most discrepancies occurred with ampicillin, penicillin G, and vancomycin, because the BIOMIC generated lower MICs. Minor discrepancies that occurred with penicillin G and staphylococci were due to Vitek programming. The Vitek reported greater than or equal to the highest well concentration for β-lactamase producers. Major discrepancies that resulted with ampicillin/sulbactam and cephalothin were also due to Vitek programming, which reports the two drugs resistant if oxacillin is resistant. Testing of clindamycin and oxacillin with staphylococci showed 100% agreement.

Table 3. Comparison of BIOMIC and Vitek System MICs for *Enterobacteriaceae*[a]

Antimicrobial agent	Enterobacteriaceae results[b]	Totals (%)
Amikacin	34, 2, 0	94, 6, 0
Ampicillin	58, 6, 7	82, 8,10
Cefamandole	29, 3, 4	81, 8,11
Cefazolin	34, 0, 1	97, 0, 3
Cefoxitin	59, 3, 9	83, 4,13
Cephalothin	30, 3, 3	83, 8, 8
Gentamicin	69, 2, 0	97, 3, 0
Nitrofurantoin	29, 0, 6	83, 0,17
Norfloxacin	34, 0, 1	97, 0, 3
Tetracycline	43,21, 7	61,29,10
Ticarcillin	65, 2, 4	91, 3, 6
Trimethoprim/ sulfamethox- azole	69, 2, 0	97, 3, 0
Totals	553,44,42	87, 7, 6[c]

[a]The sets of 3 numbers refer to results which were in agreement, minor discrepancies, and major discrepancies, respectively.

[b]n=71 isolates (36 non-urinary, 35 urinary).

[c]This set of numbers indicates the percentage of overall agreement, minor discrepancies, and major discrepancies for the 639 organism/drug combinations analyzed.

Table 4. Comparison of BIOMIC and Vitek System MICs for Nonfermentative Gram-Negative Isolates[a]

Antimicrobial agent	*Acinetobacter calcoaceticus* biotype *anitratus* results[b]	*Pseudomonas aeruginosa* results[c]	Totals (%)
Amikacin	7, 1,0	16, 9, 0	70,30,0
Aztreonam	6, 0,2	24, 0, 1	91, 0,9
Carbenicillin	7, 1,0	20, 2, 3	82, 9,9
Ceftazidime	8, 0,0	24, 0, 1	97, 0,3
Ceftriaxone	7, 0,1	17, 6, 2	73,18,9
Ciprofloxacin	8, 0,0	25, 0, 0	100, 0,0
Gentamicin	8, 0,0	22, 1, 2	91, 3,6
Imipenem	8, 0,0	25, 0, 0	100, 0,0
Piperacillin	6, 2,0	23, 1, 1	88, 9,3
Tetracycline	1, 7,0	23, 0, 2	73,21,6
Ticarcillin/ clavulanic acid	8, 0,0	20, 4, 1	85,12,3
Trimethoprim/ sulfamethoxazole	8, 0,0	24, 0, 1	97, 0,3
Totals	82,11,3	263,23,14	87, 9,4[d]

[a]The sets of 3 numbers refer to results which were in agreement, minor discrepancies, and major discrepancies, respectively.

[b]n=8 isolates.

[c]n=25 isolates.

[d]This set of numbers indicates the percentage of overall agreement, minor discrepancies, and major discrepancies for the 396 organism/drug combinations analyzed.

DISCUSSION

This study showed a 15% discrepancy rate between BIOMIC and Vitek MIC testing. Discrepancies found in this study for certain organism/drug combinations have previously been reported.[1,2,3,4,5,7,10,11] As in this study, tetracycline has been found to cause discrepancies between disk diffusion and MIC microdilution methods.[1,5,7] One explanation is that the "all-bacteria" regression line derived for tetracycline disk diffusion testing does not apply to all members of the family *Enterobacteriaceae*.[2] However, a study comparing the Vitek system to a broth microdilution MIC reference method reported major discrepancies for tetracycline testing with *Citrobacter diversus*, *Enterobacter* spp., *Klebsiella pneumoniae*, and *Serratia marcescens*.[7] Because most discrepancies were minor in this study, tetracycline testing with the BIOMIC would not be contraindicated.

The utilization of an "all-bacteria" regression line for amikacin testing with *P. aeruginosa* has also been found to be inappropriate for correlating zone diameters and MIC values.[2,3,4,10] However, the Vitek system has been shown to generate erroneous results with amikacin and other aminoglycosides for *P. aeruginosa*, when compared to a broth microdilution MIC reference method.[7,11] Because no major discrepancies occurred in this study, amikacin testing of *P. aeruginosa* with the BIOMIC system is not projected to be problematic.

The testing of nitrofurantoin with *Enterobacteriaceae* resulted in the highest percentage of major discrepancies, 17%. For an unknown reason, BIOMIC MICs were always significantly lower. However, two previous studies, which compared the BIOMIC to manual broth microdilution methods reported 97 to 100% agreement with nitrofurantoin.[5,9]

E. faecalis results exhibited very low agreement between the systems. BIOMIC MICs were consistently lower for ampicillin, penicillin G, and vancomycin. In an earlier study, which compared the Vitek system to disk diffusion, major discrepancies were found for ampicillin and enterococci.[7] Conversely, another study, which compared the BIOMIC to a manual broth microdilution method showed 98% agreement for the testing of *E. faecalis* with ampicillin and vancomycin.[5] A possible explanation for the *E. faecalis* discrepancies and those found for other organism/drug combinations in this study is the differing test methodologies. For example, different regression lines were most likely utilized for MIC calculation by the two systems.

The occurrence of minor discrepancies was expected, because of the different tests methodologies. In a clinical setting, the minor discrepancies are not projected to have had a significant impact on the section of appropriate antimicrobial therapy. In summary, this study showed the relatively inexpensive BIOMIC system to be an acceptable alternative to MIC testing with the Vitek system.

ACKNOWLEDGEMENTS

We thank Diane Halstead and Lynda Burinsky for the organisms they provided. This study was supported in part by Remel (Lenexa, KS).

Table 5. Comparison of BIOMIC and Vitek System MICs for *Enterococcus faecalis* and Staphlococci[a]

Antimicrobial agent	Enterococcus faecalis results[b]	Coagulase-negative staphylococci results[c]	Staphylococcus aureus results[d]	Totals (%)
Ampicillin	0,10,0	NT	NT	0,100, 0
Ampicillin/ sulbactam	NT[e]	10, 0,2	10, 1,0	87, 4, 9
Cephalothin	NT	10, 0,2	11, 0,0	91, 0, 9
Ciprofloxacin	NT	12, 0,0	9, 2,0	91, 9, 0
Clindamycin	NT	12, 0,0	11, 0,0	100, 0, 0
Erythromycin	7, 3,0	11, 0,1	9, 0,2	82, 9, 9
Nitrofurantoin	9, 1,0	NT	NT	90, 10, 0
Norfloxacin	10, 0,0	NT	NT	100, 0, 0
Oxacillin	NT	12, 0,0	11, 0,0	100, 0, 0
Penicillin G	3, 7,0	3, 9,0	2, 8,1	24, 73, 3
Tetracycline	10, 0,0	11, 0,1	11, 0,0	97, 0, 3
Trimethoprim/ sulfamethoxazole	NT	10, 1,1	11, 0,0	91, 4, 4
Vancomycin	1, 9,0	11, 1,0	11, 0,0	70, 3,27
Totals	40,30,0	102,11,7	96,11,3	79, 17, 3[f]

[a]The sets of 3 numbers refer to results which were in agreement, minor discrepancies, and major discrepancies, respectively.

[b]n=10 isolates.

[c]n=12 isolates.

[d]n=11 isolates.

[e]NT, Not tested.

[f]This set of numbers indicates the percentage of overall agreement, minor discrepancies, and major discrepancies for the 300 organism/ drug combinations analyzed.

REFERENCES

1. B.A. Backes, S.J. Cavalieri, J.T. Rudrick, and E.M. Britt, Rapid antimicrobial susceptibility testing of gram-negative clinical isolates with the automicrobic system, J. Clin. Microbiol. 19:744-747 (1984).
2. A.L. Barry, "The Antimicrobic Susceptibility Test, Principles and Practices", pp. 196-207, Lea and Febiger, Philadelphia (1976).
3. A.L. Barry, Procedures for testing antimicrobial agents in agar media: theoretical considerations, pp. 1-23 in: "Antibiotics in Laboratory Medicine", V. Lorian ed., Williams and Wilkins, Baltimore (1980).

4. A.L. Barry, C. Thornsberry, and R.N. Jones, Gentamicin and amikacin disk susceptibility tests with *Pseudomonas aeruginosa*: definition of minimal inhibitory concentration correlates for susceptible and resistant categories, J. Clin. Microbiol. 13:1000-1003 (1981).
5. R.F. D'Amato, L. Hochstein, J.R. Vernaleo, D.J. Cleri, A.A. Wallman, M.S. Gradus, and C. Thornsberry, Evaluation of the BIOGRAM antimicrobial susceptibility test system, J. Clin. Microbiol. 22:793-798 (1985).
6. H.M. Ericsson, and J.C. Sherris, Antibiotic Sensitivity Testing, report of an international collaborative study, Acta Pathol. Microbiol. Scand. 217(Suppl.):1-90 (1971).
7. S.L. Hansen, and P.K. Freedy, Concurrent comparability of automated systems and commerically prepared microdilution trays for susceptibility testing, J. Clin. Microbiol. 17:878-886 (1983).
8. National Committee for Clinical Laboratory Standards, Approved Standard: NCCLS Document M2-A4, Performance Standards for Antimicrobial Disk Susceptiblity Tests, 4th edition NCCLS, Villanova, PA. (1990).
9. R.L. Sautter, and G.A. Denys, Comparison of BIOGRAM and commercial microdilution antimicrobial test systems, J. Clin. Microbiol. 25:301-304 (1987).
10. J.A. Washington, P.K.W. Yu, T.L. Gavan, F.D. Schoenknecht, and C. Thornsberry, Interpretation of the disk diffusion susceptiblity test for amikacin: report of a collaberative study, Antimicrob. Agents Chemother. 15:400-407 (1979).
11. B.F. Woolfrey, B.F. R.T. Lally, M.N. Ederer, and C.O. Quall, Evaluation of the automicrobic system for susceptibility testing of *Pseudomonas aeruginosa* to gentamicin, tobramycin, and amikacin, J. Clin. Microbiol. 19:502-505 (1984).

CONTRIBUTORS

Arthur L. Barry, Ph.D.
The Clinical Microbiology Institute
Tualatin, Oregon

Raymond C. Bartlett, M.D.
Director, Division of Microbiology
Department of Pathology
Hartford Hospital
Hartford, CT 06115

Laura A. Beninsig
Baxter Diagnostics, Inc.
Microscan Division,
West Sacramento, CA

Kurt M. Vanden Brink
Baxter Diagnostics, Inc.
Microscan Division,
West Sacramento, CA

Luke J. DiMichele
Baxter Diagnostics, Inc.
Microscan Division,
West Sacramento, CA

R. Scott Evans, Ph.D.
Division of Infectious Diseases &
Medical Informatics
LDS Hospital
Salt Lake City, UT 84148

James H. Godsey, Ph.D.
Baxter Diagnostics, Inc.
Microscan Division,
West Sacramento, CA

Janet Hindler, MCLS MT (ASCP)
Technical Specialist
Clinical Microbiology Section
Department of Pathology
UCLA Medical Center
10833 Le Conte Avenue
Los Angeles, CA 90024-1713

James Jorgensen, Ph.D.
University of Texas Health Sciences Center
Department of Pathology
7703 Floyd Curl Drive
San Antonio, TX 78284

Franklin P. Koontz, Ph.D.
Director of Pathology
University of Iowa Hospitals
Iowa City, IA 52247

C. Ross McFarland, Ph.D.
Pottstown Memorial Medical Center
Pottstown, PA

Joel Mortensen, Ph.D.
Director of Clinical Microbiology
St. Christopher Hospital for Children
Erie Avenue at Front Street
Philadelphia, PA 19134

Patrick Murray, Ph.D.
Washington University School of Medicine
Department of Laboratory Medicine
660 South Euclid Avenue
St. Louis, MO 63110

W. Richard Peterson
Baxter Diagnostics, Inc.
Microscan Division,
West Sacramento, CA

Stanley L. Pestotnik, RPh
Division of Pharmacy
LDS Hospital
Salt Lake City, UT 84148

James A. Poupard, Ph.D.
Director, Clinical Microbiology
SmithKline Beecham Pharmaceuticals
King of Prussia, PA

James Prier, Ph.D.
1030 Dekalb Pike
Centre Square, PA 19422

Stephen Rittenhouse, M.A.
Clinical Microbiology
SmithKline Beecham Pharmaceuticals
King of Prussia, PA

Samuel Schalkowsky
Chief Executive Officer
Spiral System Instruments
7930 Old Georgetown Road
Bethesda, MD 20814

David G. Sherman, Ph.D.
Baxter Diagnostics, Inc.
Microscan Division,
West Sacramento, CA

Brian Slocombe
SmithKline Beecham Pharmaceuticals
Brockham Park
Betchworth, Surrey, RH3 7AJ

Christine E. Thorburn
SmithKline Beecham Pharmaceuticals
Brockham Park
Betchworth, Surrey, RH3 7AJ

Karla Tomfohrde
Baxter Healthcare Corporation
Microscan Division
1584 Enterprise Blvd.
West Sacramento, CA 95691

Lori Walsh, M.S.
Administrative Director, Pathology
Abington Memorial Hospital
Abington, PA 19001

Tammy Wolfram, M.S.
St. Joseph Hospital
Reading, PA

INDEX